MW00966224

Genetic aspects of carcass composition and meat quality in sheep

Eleni Karamichou

Genetic aspects of carcass composition and meat quality in sheep

Genetic and Genomic Aspects of Carcass Composition, Assessed by X-Ray Computer Tomography, and Meat Quality Traits in Sheep

LAP LAMBERT Academic Publishing

Impressum/Imprint (nur für Deutschland/ only for Germany)

Bibliografische Information der Deutschen Nationalbibliothek: Die Deutsche Nationalbibliothek verzeichnet diese Publikation in der Deutschen Nationalbibliografie; detaillierte bibliografische Daten sind im Internet über http://dnb.d-nb.de abrufbar.
Alle in diesem Buch genannten Marken und Produktnamen unterliegen warenzeichen-, marken- oder patentrechtlichem Schutz bzw. sind Warenzeichen oder eingetragene Warenzeichen der jeweiligen Inhaber. Die Wiedergabe von Marken, Produktnamen, Gebrauchsnamen, Handelsnamen, Warenbezeichnungen u.s.w. in diesem Werk berechtigt auch ohne besondere Kennzeichnung nicht zu der Annahme, dass solche Namen im Sinne der Warenzeichen- und Markenschutzgesetzgebung als frei zu betrachten wären und daher von jedermann benutzt werden dürften.

Coverbild: www.ingimage.com

Verlag: LAP LAMBERT Academic Publishing GmbH & Co. KG
Dudweiler Landstr. 99, 66123 Saarbrücken, Deutschland
Telefon +49 681 3720-310, Telefax +49 681 3720-3109
Email: info@lap-publishing.com

Herstellung in Deutschland:
Schaltungsdienst Lange o.H.G., Berlin
Books on Demand GmbH, Norderstedt
Reha GmbH, Saarbrücken
Amazon Distribution GmbH, Leipzig
ISBN: 978-3-8443-0073-4

Imprint (only for USA, GB)

Bibliographic information published by the Deutsche Nationalbibliothek: The Deutsche Nationalbibliothek lists this publication in the Deutsche Nationalbibliografie; detailed bibliographic data are available in the Internet at http://dnb.d-nb.de.
Any brand names and product names mentioned in this book are subject to trademark, brand or patent protection and are trademarks or registered trademarks of their respective holders. The use of brand names, product names, common names, trade names, product descriptions etc. even without a particular marking in this works is in no way to be construed to mean that such names may be regarded as unrestricted in respect of trademark and brand protection legislation and could thus be used by anyone.

Cover image: www.ingimage.com

Publisher: LAP LAMBERT Academic Publishing GmbH & Co. KG
Dudweiler Landstr. 99, 66123 Saarbrücken, Germany
Phone +49 681 3720-310, Fax +49 681 3720-3109
Email: info@lap-publishing.com

Printed in the U.S.A.
Printed in the U.K. by (see last page)
ISBN: 978-3-8443-0073-4

Contents

Acknowledgments

I am deeply indebted to my supervisor Professor Steve Bishop for guiding me with endless patience and providing me with all kind of support at all times during the course of my PhD. I have been very lucky in having such a good supervisor and with his vast knowledge, enthusiasm, and manner of scientific thinking; he has given me an admirable role-model as a researcher. At all times, I felt that I was part of a team rather than being merely supervised, and I am very grateful for his confidence. Working with him has been a great honour and I have enjoyed this time enormously. I am afraid I have not enough words of gratitude for him.

This work would not have been possible without the funding by the State Scholarship Foundation of Greece (I.K.Y.).

I have met lots of very interesting people during these years who have contributed in many aspects of my everyday life. I would also like to thank especially all of my friends, both near and far, with whom I have shared all sorts of good moments and adventures.

My sincerest and deepest thanks go to my family for their love and support. My father George, my mother Evangelia and my brother Dimitris have been always supporting me with their endless energy and determination. Without their encouragement and help I would not have coped. Special thanks to my parents for everything they have given me, this thesis is for them. My father has been constantly stimulating and, indeed, sponsoring me since my University times.

List of Publications

Refereed:

Karamichou, E., Merrell, B., Murray A., Simm, G., and Bishop, S. C. 2007. Selection for carcass quality in hill sheep measured by X-ray computer tomography, *Animal* **1**: 3-11. (Based on **Chapter 2**).

Karamichou, E., Nute, G. R., Richardson, R. I., McLean, K., and Bishop, S. C. 2006. Genetic analyses of carcass composition, as assessed by X-ray computer tomography, and meat quality traits in Scottish Blackface sheep. *Animal Science* **82**: 151-162. (Based on **Chapter 3**).

Karamichou, E., Richardson, R. I., Nute, G. R., McLean, K. A., and Bishop, S. C. 2006. A partial genome scan to map quantitative trait loci for carcass composition, as assessed by X-ray computer tomography, and meat quality traits in Scottish Blackface sheep. *Animal Science* **82**: 301-309. (Based on **Chapter 4**).

Karamichou, E., Nute, G. R., Richardson, R. I., Gibson, G. P., and Bishop, S. C. 2006. Genetic analyses and QTL detection, using a partial genome scan, for intramuscular fatty acid composition in Scottish Blackface sheep. *Journal of Animal Science* **84**: 3228-3238. (Based on **Chapter 5**).

Karamichou, E., Richardson, R. I., Nute, G. R., Wood, J.D., and Bishop, S. C. 2007. Genetic analyses of sensory characteristics and relationship with fatty acid composition in the meat from Scottish Blackface lambs. *Animal* **1**: 1524-1531 (Based on **Chapter 6**).

Chapter One
General Introduction

1.1 Introduction

In this General Introduction, the current state of the sheep breeding industry and its challenges in terms of meat quality are outlined. Additionally, the current carcass scanning methods that are used by the sheep industry, and definitions and genetic aspects of meat quality traits that are important to consumer perspective are introduced. The aim is to place the research chapters of this thesis in the appropriate methodological context, and to highlight areas that required further research before this thesis was started.

1.2 Background on the sheep meat breeding industry and its present challenges

The sheep industry worldwide faces a fundamental problem. With the exception of major lamb-exporting countries (New Zealand, UK and Australia), consumption of lamb in the developed world has markedly declined over the past 30 years (Lewis *et al.*, 1993). In order to reverse the downward trend in lamb consumption, the needs of the modern consumer have to be closely addressed (Stanford *et al.*, 1998). As outlined by Ward *et al.* (1995), consumers require meat with more lean, less fat (the minimal fat level required to maintain juiciness and flavour), consistent quality, portions that are considered good value for money, less wastage, convenience/ease in cooking and high level of choice/flexibility in available cuts. Unfortunately, the studies of Ward *et al.* (1995) have shown that lamb is currently failing to meet these consumer demands.

Before lamb carcasses can be changed to better meet consumer demand, carcasses must be evaluated using two important categories: (i) quality attributes such as tenderness, intramuscular fat, meat and fat colour, flavour; and (ii) composition attributes such as proportions of fat, lean and bone (Stanford *et al.*, 1998). Research on meat quality mainly concerns parameters relating to the eating quality and suitability for processing of meat (tenderness, colour, flavour etc.), whereas carcass quality refers to the composition (proportions of lean, fat and bone) and shape or muscularity of slaughtered animals.

The term 'meat quality' includes a variety of different aspects, the most important of which are hygiene, toxicology, nutrition, technology (function) and sensory (eating) quality. Which

3

aspect is most focused on depends on who is applying it, *e.g.* producers, industry or consumers, and in which context the concept meat quality is used. In the western countries, hygienic and toxicological meat quality generally is of a high standard, nonetheless outbreaks such as the BSE crisis greatly damage the confidence in meat and meat products. Moreover, consumers are only willing to buy meat and meat products with an acceptable eating quality; indeed, they desire meat products with visual appeal, high juiciness and high tenderness. This can sometimes be incompatible with the interests of the meat industry, which aims at optimal functional properties of the meat. One good example is the *RN* allele, present in pigs of the Hampshire breed or crossbreds with Hampshire. The *RN* allele foremost affects the water holding capacity of the meat negatively, whereas the eating quality is superior regarding tenderness and juiciness.

Moreover, factors influencing meat quality are of interest. Numerous genetic and environmental variables are involved in the final meat quality. For example meat quality can be influenced by the amount and type of fat in meat. Therefore, fatty acid (FA) composition plays an important role in the definition of meat quality, as it is related to differences in organoleptic attributes and in nutritional value of fat for human consumption (Wood and Enser, 1997). There has been an increased interest in recent years in ways to manipulate the fatty acid composition of meat. This is because meat is seen to be the major source of fat in the diet and especially of saturated fatty acids, which have been implicated in diseases associated with modern life, particularly in developed countries. These include various cancers and especially coronary heart disease. In the UK, the Department of Health (1994) recommended that fat intake be reduced to 30% of total energy intake (from 40%) with a figure of 10% of energy intake in saturated fatty acids (from 15%). At the same time, the recommended ratio of polyunsaturated fatty acids (PUFA) to saturated fatty acids (SFA) (PUFA: SFA) should be increased to above 0.4. Since some meats naturally have a PUFA: SFA ratio of around 0.1, meat has been implicated in causing the imbalanced fatty acid intake of today's consumers (Wood *et al.*, 2004). Also, the ratio of omega-6: omega-3 fatty acids is a risk factor in cancers and coronary heart disease (Enser, 2001). The recommendation is for a ratio of less than 4 and again some meats are higher than this. As with the PUFA: SFA ratio, meats can be manipulated towards a more favourable omega-6: omega-3 ratio. The ratio of omega-6: omega-3 PUFA is particularly beneficial (low) in ruminant meats. Ruminants also naturally produce conjugated linoleic acids (CLAs) that may have a range of nutritional benefits in the diet (Enser, 2001).

Simultaneously, prediction of carcass composition by the proportions of the fat, lean and bone components of the carcass in live animals, as well as the subsequent fat and conformation grading the carcass receives in the abattoir, has an important use in evaluating meat animals. Lean is the most valuable part of the carcass and most sought after by consumers. Consequently, prediction of its content in live animals has significant application to the meat industry. Several techniques have been used to predict carcass composition in live animals, such as computer tomography (CT) scanning, ultrasound and nuclear magnetic resonance. Carcass fat content affects the overall value and productivity of a sheep production enterprise in several ways. For example, the fatness of lamb carcasses at slaughter affects the overall efficiency of lean meat production, as well as the perceived desirability of the meat (Bishop *et al.*, 1995).

The above aims of improving carcass composition and meat quality can be realised by means of genetics, which potentially enables cheap, cumulative, permanent and sustainable improvements within production systems. For many years breeders have changed the genetic make up of animals through selection without knowledge of the underlying genes. Until recently, the tools to identify the genes responsible for genetic differences between individuals or between populations have not been available. Developments in the area of molecular biology have changed this situation and have allowed genes or chromosomal segments containing genes of importance to be identified in humans and farm animals (Bovenhuis *et al.*, 1997).

Current approaches in animal breeding, to estimate the genetic value of an individual, depend upon phenotypic observations on the individual itself and/or relatives. For almost all the traits of interest to animal breeders, differences in phenotypic observations are due to both genetic and environmental differences (Bovenhuis *et al.*, 1997). Furthermore, segregation of alleles takes place each time genes are transmitted from parent to offspring. As a result of these two factors, accurate estimation of the breeding value of an animal is possible only if large numbers of records are available. If genes and their effects on traits of interest are known, typing of animals at the DNA level may make it possible to estimate breeding values independent of phenotypic observations. A key approach for utilising genetics in this way is the identification of genetic markers associated with the phenotypic traits of interest. This is especially for the traits of the greatest direct relevance to the sheep sector, *viz* the composition of the carcass and the quality of the subsequent meat. Genetic markers would both enhance this technology and reduce its expense, possibly making it more accessible to a greater number of breeders. Meat quality traits currently cannot be directly measured on animals participating in breeding programmes and, hence, cannot be easily included in breeding programmes. The use of genetic markers is a possible solution to this problem. Therefore, genetic markers for meat quality, including aspects of fatty acid composition, may allow the breeding of sheep with meat of a higher quality for the consumer.

1.3   An overview of carcass composition

As described above, carcass composition is important in terms of carcass value, consumer appeal and production efficiency. Therefore, accurate carcass evaluations and selecting good carcasses or meat cuts to fulfil market demand is becoming more and more important. Carcass weight and meat grading are no longer adequate to accurately evaluate products. At the present time research is concentrating on more sophisticated processing technology. The following section concentrates on these principles in carcass composition and on various methods used to estimate such composition.

1.3.1   Growth, development and body form

Young *et al.* (1996) suggested a model of body form that accounts for all major variations in carcasses. The major determinants of body form can be separated into five main categories:

1. Size overall (*e.g.* bodyweight)
2. Proportions of major tissues (*e.g.* fat %, muscle to bone ratio (M: B))
3. Distribution and partitioning of tissues (*e.g.* % muscle in high priced cuts, % muscle in loin, % fat in subcutaneous depot)
4. Shape of tissue units *e.g.* muscularity (*e.g.* volume relative to length, width relative to depth)
5. Density of tissues (weight relative to volume, chemical composition).

Animal weight (category 1) is simply a function of the volume (3 dimensional images) and density (2 dimensional images) of the three major tissues, their distribution and shape. However, since animals can be selected for slaughter at different stages of growth, a real interest is on the form of animals of a similar size and on how these other traits change with overall size during growth.

It is reasonable to expect that changes or differences in one factor are associated with changes in one or more of the others, *e.g.* during growth the proportion of muscle changes

6

relatively little whereas decreases in % bone are offset by increases in % fat (Young *et al.*, 1996). This will affect body form both directly (category 2) and through changes in shape of tissue units (category 4) concomitant with changes in the relative size of muscles compared to the bones they are associated with (Young and Sykes, 1987).

Value of carcasses is affected by categories 1 and 2 of this model through carcass size and fatness, and commercial carcass classification systems focus on these two characteristics (Kempster, *et al.*, 1982; Price, 1995). To what extent other components are valuable is less clear. Many classification systems assess "conformation" which is related to shape of the carcass overall (Price, 1995). This is a less than useful definition as it has many different interpretations and it is almost always associated with fatness within a genotype (Kempster *et al.*, 1982; Kirton *et al.*, 1983; Jones, 1995). More useful is "muscularity" which refers to the shape of muscles. "Blocky" (thick relative to length) is the desired shape (Butterfield, 1988).

As animals grow, fatness increases and muscles become blockier (Young and Sykes, 1987; Butterfield, 1988; Price, 1995; Abdullah *et al.*, 1998). Since carcass bone proportion decreases, muscle to bone ratio (M: B) increases as well (Young, 1989). Traditionally this "fleshing out" of the carcass as animals grew was highly desirable. However, with the general tendency against fat, traditional values for the lean carcass (blocky muscles and high M: B) are compromised since the most obvious way to reduce fat is select animals that are less mature. That this has occurred is borne out by the results of many breeding programmes that have reduced fatness by selecting animals that are less mature at a given age or size (Lewis *et al.*, 1996; Abdullah *et al.*, 1998; Emmans, *et al.*, 2000). Thus animals are larger but less mature, fitting neatly the theory of genetic size scaling described by Taylor (1985).

1.3.2 Methods of non-invasive imaging techniques to predict carcass composition

Methods of estimating carcass composition are comprised of two types. They either describe the chemical composition of the body or the anatomical distribution of its tissues. In each case the techniques used can be either destructive or non-destructive to the animal. Non-destructive techniques are frequently required, *e.g.* studies with humans, studies with valuable animals, and when sequential studies of the same subject are required (Szabo *et al.*, 1999).

Imaging techniques produce an image of a body component that is then used to predict chemical or physical body composition. Therefore, there are two main factors affecting the accuracy of estimation: the correlation between image and body part, and the correlation between body part and total body composition (tissular or chemical) (Szabo *et al.*, 1999).

In this section some of the newer methods, which can produce cross section images of the body (real time ultrasound imaging, magnetic resonance imaging (MRI) and computer tomography (CT)) will be reviewed to evaluate their applicability in animal breeding.

1.3.2.1 Real-time ultrasound imaging

The first publications using ultrasound to measure backfat thickness date back to the 1950s (Temple *et al.*, 1956; Dumont, 1957; East *et al.*, 1959). Further development of techniques enables researchers to visualize the cross-section view of most or all near-skin tissues.

The principle of the ultrasound procedure is that high-frequency (1–5 MHz) sound signals pass through the body tissues, and when an interface (boundary) between two tissues is encountered, some sound waves are reflected. The reflected signals are detected by the receiver and can be amplified and shown in visual form. Real-time ultrasound imaging equipment emits many sound waves simultaneously using a specially designed transducer along a linear path. The reflected sound waves are then transcribed and presented on a monitor as a two-dimensional image or cross-section of muscle mass, such as, for example, the loin-eye muscle (Szabo *et al.*, 1999). The disadvantage of this equipment is that the sound signals do not go deep enough in the body to scan the whole cross-section of the body.

Real-time ultrasound imaging is capable of detecting small areas of intramuscular fat or marbling. However the quantitative determination of these is more difficult. Technicians cannot differentiate visually between signals reflected by intramuscular fat and those reflected by connective tissue, blood veins and arteries, which permeate muscle tissue (Miller, 1996).

Prediction accuracies, reported by Alliston (1983) and Simm (1987), for prediction of carcass tissue weights from live weight and ultrasound tissue depths in sheep, ranged from 50 to 70%. In addition, the study by Young *et al.* (1996) in Dorset Down lambs, reported an accuracy of up to 83% for lean weight and 63% for fat weight. In Scottish Blackface lambs, Bishop (1994) reported an accuracy of 73% for lean weight and 80% for fat weight.

1.3.2.2   Magnetic resonance imaging (MRI)

The principle during the MR examinations is that the atoms of the body, positioned in a strong magnetic field, absorb energy from an external energising source and then emit radio signals, as a function of time, which can be computer-processed to form an image (Szabo *et al.*, 1999). This phenomenon is called nuclear magnetic resonance (NMR).

MR instruments for animal investigations are available at a very few sites. Thus, limited information is available in the literature. Most of the published studies (Fowler *et al.*, 1992; Hennig, 1992; Scholz *et al.*, 1993; Baulain, 1997) reported that the proportion of tissues is predicted with lower accuracy than that achieved for absolute mass. Most of the R^2 values obtained for tissue mass are very high (80 - 90%).

In the sole study where MRI has been used to evaluate body composition in sheep, Streitz *et al.* (1995) reported R^2 values ranging from 78 to 91% for percentage of lean in lambs at body weights from 10 to 50kg. Presently, programs using MRI for the genetic improvement of carcass quality in livestock are restricted to poultry (Mitchell *et al.*, 1991; Liu *et al.*, 1994). Additional studies with sheep would be required before the benefits of using MRI for prediction of body composition could be evaluated relative to costs.

1.3.2.3   Computer tomography (CT)

Computer tomography had already been used in humans for many years for diagnostic purposes when it was realized that it might be a valuable tool for the estimation of the body composition of animals. The first report on the use of CT in animal science was by Skjervold *et al.* (1981). Early research highlighted the potential of CT scanning for the study and evaluation of animals (Vangen and Skjervold, 1981; Vangen, 1988; Young *et al.*, 1987) and theoretical predictions presented a clear and convincing case for supplementation of ultrasound scanning with CT scanning of elite animals in breeding programmes. Such predictions forecast increases in genetic progress that may be as high as 50% (Simm and Dingwall, 1989) or closer to 100% (Jopson *et al.*, 1995) through use of CT in combination with ultrasound scanning.

CT is based on the differential that exists between the rates at which the tissues in the body attenuate X-rays. Attenuation measures are collected by detectors, which rotate 360° around the body in synchrony with the X-ray source. The data are then processed to produce a matrix of CT values for the target cross-section of the body. The matrix of CT numbers is converted to different grey values and then displayed as an image on a monitor. The image can be stored digitally and used later for evaluation or for analysis to generate prediction equations for carcass composition traits. CT numbers are in close correlation with the density of tissues. Values are expressed in Hounsfield units (Hu) (Szabo *et al.*, 1999).

CT allows cross-sectional images, containing a wealth of information, to be obtained for a living animal (Young *et al.*, 1996). These can be used to provide very accurate assessment of body composition in live animals in a welfare-friendly manner. Not only is accuracy improved, compared to ultrasound, but also a wide range of novel traits lend themselves to assessment, and objective measurements can be collected rapidly (Young *et al.*, 1996).

Initially, CT was used to measure body composition in animals at different live weights. Most of this early work was based upon the use of prediction equations to estimate body composition. Vangen and Baulain (unpubl.) and Kolstad and Vangen (unpubl.) used prediction equations developed for pigs of normal slaughter weights to estimate the weight of protein and fat in 25, 55 and 95kg pigs. They demonstrated that it was possible to measure differences between sexes, breeds and feeding intensities in the above different weights. Subtle but significant differences at the various weights were found between the two sexes (gilt versus barrow) and between the various genetic groups (proportion of Duroc genotype: 0, 25 or 50%). The use of CT to estimate body composition had therefore given new information about composition traits. Based on this data it was possible to quantify fat and protein growth and consequently estimate real biological efficiency.

Prediction equations have been shown to be very accurate for the group of animals for which each was developed. Sehested (1984) reported that CT values with live weight could predict kg fat-free lean lamb(s) with R^2 values of 92 to 94%, RSD 0.5 to 0.6kg, compared to prediction with live weight alone (R^2= 0.83, RSD= 0.8kg). Those values were in agreement with the 92 to 98% found more recently by Vangen and Jopson (1996). In Dorset Down ewe lambs, R^2 values of 93%, 94% and 73% for fat, lean and bone,

10

respectively, were found using four CT scans and live weight to predict the tissue weights (Young *et al.*, 1996).

A later study by Young *et al.* (1999) demonstrated the value of CT in UK sheep. Table 1 summarises accuracies of prediction from a number of trials. In that study Young *et al.* (1999) showed that, by using live weight and tissue areas from three scans, prediction accuracy (R^2) of 97%, 99% and 89% could be achieved for the total weight of muscle, fat and bone, respectively, in the carcass of Suffolk lambs. In addition, Young *et al.* (1999) showed that the total weight of muscle in the carcass of Suffolk lambs could be predicted accurately (R^2= 96%) using equations that included live weight and the area of muscle on CT scans from the 8th thoracic vertebrae, the 5th lumbar vertebrae (LV5) and the ischium (ISC) positions. A consequence of the increased accuracies shown for the CT results is high heritabilities (0.40 - 0.50) for fat weight and lean weight; one of the reasons CT scan data helps the breeders.

Similar breed-specific equations for total weight of muscle in the carcass, which included the same predictors, were also developed for Charollais and Texel sheep, with prediction accuracies (R^2) of 98 and 96% respectively (M. J. Young *et al.*, unpublished). For Scottish Blackface ewes, Lambe *et al.* (2003) showed that fat and muscle weights were predicted accurately from CT (R^2= 80 to 99%) but bone weight was predicted less accurately (R^2= 56%). The results of Lambe's *et al.* (2003) study suggested that computed tomography would allow accurate within-animal predictions of fat (in various depots), muscle and bone levels, which can be measured throughout the lifetime of breeding ewes.

Table 1.1 *Prediction accuracy of CT scanning (R^2, residual standard deviation and mean). Predictors are carcass tissue areas from two or three scans plus live weight*

Carcass tissue	Fat	Muscle	Bone
Prediction of Meat Sheep	**Weight (g)**	**Weight (g)**	**Weight (g)**
R^2 (%)	99	97	89
RSD (g)	434	611	313
Predicted variable mean (g)	8620	13880	4130
Hill lambs			
R^2 (%)	92	86	73
RSD (g)	191	388	184
Predicted variable mean (g)	2820	7930	2550

Young *et al.*, 1999.

1.3.3 Strengths and opportunities with *in vivo* estimation of carcass composition

The keys to changing carcass composition to better meet consumer demand are methods of evaluating body composition *in vivo*. For example, computer tomography (CT) technology offers great possibilities for studying body composition in the live animal. Such technology can be utilised in genetic studies with the aim of quantifying short or long-term changes in body composition and tissue depots. Body tissues can be measured down to individual depots or muscles. The strength of these techniques is that the between animal variation normally associated with a slaughter trial is removed. If body composition measurements are taken over time, the changes in composition can be analysed on a within- animal basis. Additionally, body composition measurements can be made in high value or rare animals without the need to slaughter them.

Ultrasound scanning has been used widely in sheep breeding programmes to address the problem of over-fatness in lamb by genetically improving the rate of lean tissue growth (Simm, 1994; Stanford *et al.*, 1998). X-ray CT scanning can provide more accurate information on body composition, *in vivo*, in sheep than ultrasound (Sehested, 1984; Young *et al.*, 1996, 2001). Thus, CT has the potential to improve rates of genetic gain from selection by as much as 50% when used in combination with ultrasound scanning (Simm and Dingwall, 1989; Jopson *et al.*, 1995). Opportunities offered by CT have already been incorporated into sheep breeding programmes in New Zealand, the UK and Norway, where information from CT has been included in genetic evaluations (Nicoll *et al.*, 1997; Young *et al.*, 2001; Vangen *et al.*, 2003).

Consequently, CT scanning has solved many of the difficulties of measuring short-term changes in body composition. This technique offers considerable benefits to quantitative genetics. Probably the most significant of these benefits is that adult traits can be measured in the same animals in which carcass composition was measured, at what would have been slaughter age. For example, CT could provide information on other traits affecting carcass and meat quality that cannot be measured easily by other means on the live animal. The genetic relationships between carcass and meat quality traits could be more accurately calculated. To date, there have been no published examples of this usage of CT and this thesis aims to investigate these relationships.

1.4 Meat Quality

Meat quality has risen to the forefront of consumer demand, a trend that is likely to continue. Therefore, it becomes necessary to understand and investigate the factors influencing meat quality. The complex concept of meat quality can be divided into production quality and product quality (Hofmann, 1994). Product quality can be further divided into technological (*i.e.* the functional properties of the meat), sensory (eating experience), nutritional (chemical composition, healthiness of meat) and toxicological/hygienic (absence of *e.g.* contaminants, harmful micro-organisms) quality. As both visual meat quality and sensory attributes of the meat can generally only be directly measured in the slaughtered animal, there is limited opportunity to include those traits in a breeding programme, without either a complex breeding programme design or the use of genetic markers or some other *in vivo* predictor. Therefore, this thesis focuses on quantitative genetic aspects of technological, sensory and nutritional meat quality.

This section will start by examining the biology and properties of various meat quality attributes (*i.e.* colour, flavour, juiciness, and intramuscular fatty acids) and genetic aspects of these traits will be considered later in this section.

1.4.1 Technological meat quality

The technological quality describes the suitability of the raw material for further processing as well as the yield during processing. The technological quality attributes of meat include its water-holding capacity (WHC), colour, texture (tenderness) and chemical composition. These parameters are influenced by multiple interacting factors including breed, genetics, feeding, pre-slaughter treatment, slaughter method, chilling and storage conditions as reviewed by Rosenvold and Andersen (2003). The occurrence of a few known genes, especially the halothane (a point mutation in the ryanodine receptor, Fujii *et al.*, 1991) and RN⁻ genes, considerably affects the quality characteristics of fresh and processed pork. However, when major gene effects are excluded, genetics explains less of the variation in meat quality traits, which are often only slightly, to moderately heritable (Sellier, 1998). There is a large variation in meat quality both within and between animals of the same breed, sex and environment, which is not well understood. This variation is likely to be caused by differences in various known and unknown intrinsic (genetics) and extrinsic (environmental) factors, which interact and determine the outcome of metabolic processes in the pre- and post-mortem period (Klont *et al.*, 1998).

At slaughter of the animal, the supply of oxygen, glucose and free fatty acids to muscle ceases when the blood circulatory system stops. Any subsequent metabolism must be anaerobic and adenosine triphosphate (ATP) can only be regenerated through breakdown of glycogen through glycolysis, since oxidative decarboxylation and phosphorylation no longer operate. Lactic acid accumulates and the muscle gradually acidifies. The decline in pH depends on the initial concentration of creatine phosphate and glycogen (Bendall, 1951) and the characteristics of the post-mortem pH decline are determined by the physiological condition of the animal at stunning (Bendall, 1973; Warriss, *et al.*, 1989). For example, in a well-fed and unstressed pig, the pH value typically falls from about 7.2 to an ultimate pH (pHu) of about 5.4, reaching a plateau when enzymes participating in the glycolysis are inactivated (Lawrie, 1992). The biochemical and physical processes taking place during the post-mortem conversion of muscle to meat are crucial for the final product; more specifically, post-mortem pH and temperature development are very important causes of variation in pork quality (Sellier and Monin, 1994; Schäfer *et al.*, 2002). Both the rate and extent of pH decline are of importance (Briskey, 1964). The rate of pH decline is often indicated by values measured 45 minutes post-mortem (denoted pH$_{45}$) whereas the extent (pH$_u$) is normally measured 24 h post-mortem but may decline further. In a study by Josell *et al.* (2003a) pH$_u$ values were not reached until 48h post-mortem.

1.4.1.1 Colour

Meat colour is an important quality attribute for the consumer (Risvik, 1994). As for WHC, the temperature and pH history of the muscle post-mortem are of importance for the meat colour through their effects on the physical structure and light scattering properties of the meat proteins (MacDougall, 1982). WHC can be defined as the ability of meat to retain inherent water during storage, processing and cooking. Water loss, and subsequent inferior technological quality, causes financial losses for the industry and results in a less attractive product appearance for the consumer. WHC may also influence the sensory quality of meat. Further, meat colour is affected by the concentration and properties of the meat pigments myoglobin and, to a lesser extent, haemoglobin (Lindahl *et al.*, 2001). Muscle myoglobin (80 - 90% of total pigments) concentration varies between species, breed, sex, age, type of muscle and level of training (Ledward, 1992). In fresh meat myoglobin can exist in three different forms: the reduced form of myoglobin (deoxymyoglobin) is purplish, and the oxygenated form (oxymyoglobin) is bright red whereas the oxidized form (metmyoglobin) is brown. Fresh meat colour is affected by the relative abundance of these three forms (Govindarajan, 1973).

Both the light scattering and WHC phenomena can be illustrated by two quality extremes; pale soft and exudative (PSE) and dark firm and dry (DFD) meat. PSE is associated with pale meat with low water-holding capacity. It is caused by extensive protein denaturation and lateral shrinkage of the myofibrils due to an early post-mortem pH decline in muscles with a still relatively high temperature (Bendall and Wismer-Pedersen, 1962; Briskey, 1964; Offer and Knight, 1988a). In PSE meat, light does not penetrate far into the meat but is scattered, which makes the meat appear pale (Offer *et al.*, 1989). PSE develops for several reasons; the most investigated is pre-slaughter stress, often in stress-sensitive carriers of the Halothane gene, increasing the post-mortem glycolytic rate and muscle temperature (Gariepy *et al.*, 1989).

DFD meat, on the other hand, is characterised by high pH_u, dark appearance and high WHC. The pH (> 6.0) is much higher than the isoelectric point of actomyosin, resulting in more water retained between the myofilaments and less exuded out of the meat matrix (Offer *et al.*, 1989). The DFD muscle has a compact structure and appears darker because its surface only slightly scatters incident light (Govindarajan, 1973). DFD is caused by low muscle glycogen levels at slaughter resulting in restricted formation of lactate *post mortem* and a rapid onset of *rigor mortis*. Low glycogen levels in turn, may be due to long-term stress, lack of food for several days or strenuous exercise (Bendall and Swatland, 1988; Warriss *et al.*, 1989).

15

1.4.2 Sensory quality of meat

Even within the field of sensory science the statement 'meat quality' can be interpreted in several ways. Jul and Zeuthen (1981) defined meat quality as the 'total degree of satisfaction the meat gives to the consumer'. The three sensory properties by which consumers most readily judge meat quality are texture (tenderness), juiciness and flavour of meat (Liu *et al.*, 1995), and these are discussed in this section. Meats are normally highly desired for their distinctive and highly prized flavours. However, departures from normal flavours are not uncommon in meat products and result in poor acceptability or even rejection by consumers (Buckley *et al.*, 1996).

The sensory tests available can be divided into two main groups, objective and subjective tests. The analysis of objective tests all concern product properties, whereas the subjective tests concern the consumers' opinion (preference/liking) of the product.

1.4.2.1   Texture

Several studies show that tenderness is one of the major attributes, the perceived sensory quality of meat (Wood, *et al.*, 1992; Van Oeckel *et al.*, 1999). Touraille (1992) showed that 78% of French consumers considered tenderness as very important, compared with 83% and 77% for taste and odour. Tenderness of the *muscle longissimus thoracis et lumborum* (LTL) is of special interest, because this is the part of the carcass usually destined for fresh consumption (Oeckel *et al.*, 1999). Additionally, it has a major effect on eating quality, along with pH and meat colour, which are also important as they affect keeping quality and visual appeal (Hopkins and Fogarty, 1998).

Two methods are used for evaluation of tenderness. One is by a taste panel, which is time consuming and costly, and the other one is by instrumental Warner- Bratzler shear force, which is often used as a measure for meat tenderness (Boccard *et al.*, 1981). Warner-Bratzler devices and sample configuration vary considerably; however, the general principle is to measure the force needed to cut through a standardized meat sample. Using the recommended equipment, a force deformation curve is obtained from which the parameters peak force and total work are measured (Honikel, 1998). Two major phenomena are known to influence final tenderness in meat: (i) shortening of muscle fibres (contraction during rigor) affects meat toughness and (ii) ageing affects the gradual tenderisation. Post-mortem degradation of myofibrillar proteins is the main reason for the improvement in meat tenderness during ageing (Quali, 1992). Also, for the processes of tenderisation, a combination of pH and temperature decline has been found to be important, through a possible effect on proteolytic enzymes during rigor development (O'Halloran *et al.*, 1997). Furthermore, the level of intramuscular fat (IMF; Bejerholm and Barton-Gade, 1986; van Laack *et al.*, 2001) as well as the live animal growth rate and protein turnover (Kristensen *et al.*, 2002) may affect meat tenderness and shear force values. The properties of intramuscular collagen may also influence shear force values. However, for pigs that are slaughtered young the immature connective tissue does not significantly influence pork tenderness (Avery *et al.*, 1996). These suggested factors affecting tenderness are, as reviewed by Wood *et al.* (1992), in turn affected by several processing factors, including pre-slaughter handling, stunning method, carcass chilling rate, carcass suspension system, and ageing time. Processing factors are generally more important for pork tenderness than production factors such as breed, carcass weight, fat level and feeding regime (Wood *et al.*, 1996).

1.4.2.2   Juiciness

Only a few studies have been made to obtain a basic knowledge of factors of importance to juiciness, even though juiciness facilitates the chewing process as well as bringing the flavour component in contact with the taste buds. Juiciness is therefore of great importance for the overall eating experience and should not be overlooked as an important eating quality attribute in meat (Aaslyng *et al.*, 2003).

Juiciness is the feeling of moisture in the mouth during chewing. It is a combination of moisture chewed out of the meat and saliva production mixed into the meat. The juiciness of meat depends on the raw meat quality and on the cooking procedure. Eikelenboom *et al.* (1996) showed that juiciness is slightly correlated to intramuscular fat (IMF) (r=0.33) but even more correlated to ultimate pH_u (r=0.68). Dransfield *et al.* (1985) however, found a quadratic relationship between juiciness and pH_u with a minimum at pH_u =6.1, but the pH_u only explained 5% of the variation in juiciness. The water holding capacity (WHC) of pork might also influence the juiciness independent of pH_u (Aaslyng *et al.*, 2003). Other factors like concentration of glycogen could also influence the juiciness, as an increased concentration of glycogen increases the juiciness in beef with a normal pH (between 5.5 and 5.75) (Immonen *et al.*, 2000). Rearing conditions may also influence the juiciness as meat from indoor reared pigs has been shown to be juicier than meat from pigs reared outdoor (Jonsäll *et al.*, 2001). The reason for this is not known.

1.4.2.3   Flavour

Flavour is another very important component of the eating quality of meat and there has been much research aimed at understanding the chemistry of meat flavour and at determining those factors during production and processing of meat which influence flavour quality (Mottram, 1998). Consumers consider flavour to be one of the most important sensory traits of meat, and the absence of off-flavours is expected to be critical for acceptance (Risvik, 1994).

The characteristic flavour of cooked meat derives from thermally induced reactions occurring during heating, principally the Maillard reaction between amino acids and reducing sugars, and the degradation of lipid. Both types of reaction involve complex reaction pathways leading to a wide range of products, which account for the large number of volatile compounds found in cooked meat.

Lipid degradation provides compounds which give fatty aromas to cooked meat, and compounds which determine some of the aroma differences between meats from different species. Compounds formed during the Maillard reaction may also react with other components of meat, adding to the complexity of the profile of aroma compounds. For example, aldehydes and other carbonyls formed during lipid oxidation have been shown to react readily with Maillard intermediates. Such interactions give rise to additional aroma compounds, but they also modify the overall profile of compounds contributing to meat flavour. In particular, such interactions may control the formation of sulphur compounds, and other Maillard-derived volatiles, at levels which give the optimum cooked meat flavour characteristics (Mottram, 1998).

The flavour of meat is dependent upon factors such as the animal's age, breed, sex, nutritional status, differences in fat content and manner of cooking. Most important to the final flavour of the meat is relative length of time in the post-mortem ageing process, as it is during this time that many chemical flavour components are formed (Spanier *et al.*, 1990).

1.4.3 Healthiness of meat

Meat has been criticised on health grounds because of high levels of saturated fatty acids (SFA) and trans-monounsaturated fatty acids (MUFA) presumed to increase the risk for the development of coronary heart disease. Conversely, polyunsaturated fatty acids

(PUFA), which lower blood cholesterol concentrations, are often present at low levels in meat, especially those of the omega-3 family (C18:3n-3 and its derivatives) and of the omega-6 family (C18:2n-6 and its derivatives) which have particularly beneficial effects on health. These fatty acids have to be provided by diet. Linoleic acid (C18:2n-6) is essential for growth and reproduction. Linolenic acid (C18:3n-3) is essential for brain and retina functions. Moreover, omega-3 PUFA exerts a positive influence on the prevention of cardiovascular diseases (De Lorgeril et al., 1994). A too high value of the ratio of omega-6 to omega-3 PUFA is associated with an increased risk of atherosclerosis or coronary diseases. Generally, the recommended average value for this ratio for human nutrition is 2 (Okuyama et al., 1997). In this aspect, ruminant meat (bovine and ovine) is superior to pork, since its values for the ratio omega-6 to omega-3 PUFA are between 1 to 2, versus 7 for pork (Wood et al., 1999).

Thus, the perceived 'healthiness' of meat is becoming a key quality issue for consumers and this is largely related to its fat content and its fatty acid composition.
Fatty acids can be classified into the following three groups:

- **Saturated fatty acids** (SFA) contain carbon atoms linked only by single bonds and are usually solid at room temperature. They are principally obtained from animal fats and animal products (e.g. meat fat, milk, butter, cheese and cream). The most common dietary SFA are palmitic and stearic acid. These are important for energy metabolism, cell membrane structure and normal growth. However, chronic excessive intake or synthesis, or both, of palmitic and stearic acids tends to raise the level of low density lipoprotein (LDL) cholesterol ('bad' cholesterol) in blood. A high intake, therefore, could enhance the process of atherogenesis and increase the risk of cardiovascular disease (CVD) (Mason, 2004).

- **Monounsaturated fatty acids** (MUFA) contain only one double bond and are usually liquid at room temperature. The most concentrated dietary sources of MUFA are olive oil and rapeseed oil. However, MUFA also comprise about one third of the fatty acid content of meat fat. The main dietary MUFA is oleic acid. There is enormous interest in the health implications of MUFA mainly because, unlike saturates, they do not raise blood cholesterol. Substituting SFA with MUFA, therefore, lowers LDL cholesterol (Mason, 2004).

- **Polyunsaturated fatty acids** (PUFA) contain two or more double bonds and are liquid at room temperature. They can be divided into two types: n-6 (or omega-6) and n-3 (or omega-3) which have different metabolic effects. The parent fatty acid in each of these groups, linoleic acid (C18:2n-6) and alpha-linolenic acid (C18:3n-3) are also called "essential fatty acids" (EFA) because humans lack the enzymes to make them and must, therefore, get them from the diet. PUFA have a crucial role in many processes. For example, they act as components of phospholipids in cell membranes, regulators of cholesterol metabolism and precursors of eicosanoids (*e.g.* prostaglandins, leukotrienes, thromboxanes) (Mason, 2004). Eicosapentaenoic (EPA) and Docosahexaenoic acid (DHA) have a structural role in cerebral, retinal and nervous tissue and play an important part in neural developments in fetal and early life (Mason, 2004).

The omega-6 PUFA have a hypocholesterolaemic effect and their substitution for SFA has been encouraged since the 1970s – the butter *vs* margarine debate. However, as well as lowering LDL cholesterol, omega-6 PUFA also lower "good" high density lipoprotein (HDL) cholesterol. PUFA are also susceptible to oxidations and high intakes can lead to excessive free radical production and potential adverse effects (*e.g.*, development of atherosclerosis and cancer). Unlike omega-6 PUFA, MUFA do not lower HDL cholesterol and are associated with less risk of lipid per-oxidation, which means less free radical production. Furthermore, unlike omega-6 PUFA, omega-3 have a minimal effect on blood cholesterol levels, although in doses exceeding 1g per day they can reduce triglyceride concentrations (Mason, 2004).

1.4.3.1   Intramuscular fat and fatty acid composition

Intramuscular fat or more correctly, lipid content of meat, can vary substantially in *m. longissimus dorsi* (LD) and *m. biceps femoris* (BF) typically in the range of 0.5 - 4% (Essén-Gustavsson *et al.*, 1994; Fernandez *et al.*, 1999). Traditionally, lipids have been divided into structural lipids and depot lipids. Intramuscular lipids consist mainly of neutral lipids (triacylglycerols) and phospholipids and they function as vehicles for energy storage (depot), as well as being essential parts of cell membranes (structural). In addition, lipids form the basis of steroid hormones as precursors in eicosanoid metabolism. They are a highly concentrated storage form for energy and their energy value is almost double that of carbohydrate and protein.

In recent years, there has been an increased interest in producing meat with improved fatty acid composition, designed to meet the dietary recommendations for humans, as mentioned in section 1.2. The driving force has been to increase the omega-3 fatty acids, and thus improve the omega–6 to omega-3 ratio. This is due to the fact that linolenic acid (C18:3n-3), abundant in fresh forages (> 50% total fatty acids) (Bauchart *et al.*, 1984), is stored in significant amounts in ruminant tissues (Enser *et al.*, 1999). Although an important proportion of linolenic acid is converted to its saturated counterpart (stearic acid, C18:0) by ruminant biohydrogenation, small but significant amounts escape ruminant metabolism (Bauchart *et al.*, 1984, 1996) and are subsequently absorbed by the small intestine (Wood and Enser, 1997). Therefore, it is possible to induce a modification of the fatty acid composition of tissue lipids.

A survey conducted by Enser *et al.* (1996) illustrated the differences in fatty acid composition and content between beef, lamb and pork. Fifty loin steaks or chops from each species were purchased from four supermarkets to represent the meat on sale to the public. The total fat content of the steaks (obtained by dissection) was highest in lamb, probably because of a lower level of fat trimming during butchery. The total fatty acid composition of the longissimus muscle, including some fat attached to the perimysium, was also highest in lamb and was least in pork. The most obvious difference in fatty acid composition was that linoleic acid (C18: 2n-6) was higher in pork, causing a higher PUFA: SFA ratio. This is due to the high content of linoleic acid in the cereal-based diets consumed by pigs and this produced an undesirably high omega-6: omega-3 ratio.

1.4.3.2   Lipid oxidation – its role in meat quality

The scientific literature pertaining to lipid oxidation in meat and other biological systems is vast, yet it is not complete. The data are often contradictory and there remain unanswered questions. What is well established, however, is that lipid oxidation is a major deterioration reaction which often results in a significant loss of meat product quality. It is also well known that lipid oxidation is positively correlated with pigment oxidation, but the basis for this relationship is not fully understood (Liu *et al.*, 1995). Fatty acid composition also affects shelf-life of the meat product. This is explained by the propensity of unsaturated fatty acids to oxidise, leading to the development of rancidity as display times increase. From the viewpoint of meat colour, it may be that radical species produced during lipid oxidation act directly to promote pigment oxidation and/or indirectly by damaging pigment-reducing systems (Liu *et al.*, 1995). In addition to adverse changes in the colour, flavour and texture of meats (Kanner, 1994), the autoxidation of unsaturated lipids and cholesterol, results in the generation of potentially toxic compounds (Addis and Park,

1989). Antioxidants, especially α-tocopherol (vitamin E) have been used to delay lipid and colour oxidation and to extend shelf life.

The development of oxidative off-flavours (rancidity) has long been recognized as a serious problem during the holding or storage of meat products. Rancidity in meat begins to develop soon after death and continues to increase in intensity until the meat product becomes unacceptable to consumers. The biochemical changes that accompany post-slaughter metabolism and post-mortem ageing in the conversion of muscle to meat give rise to conditions whereby the process of lipid oxidation is no longer tightly controlled and the balance of prooxidative factors/antioxidant capacity favours oxidation. The propensity of meats and meat products to undergo oxidation depends on several factors including pre-slaughter events such as stress and post-slaughter events such as early post-mortem pH, carcass temperature, cold shortening, and techniques such as electrical stimulation (Buckley *et al.*, 1989). Furthermore, any disruption to the integrity of muscle membranes by mechanical deboning, grinding, restructuring or cooking alters cellular compartmentalization.

1.4.4 Genetics of Meat Quality

As we consider the genetic improvement of meat quality it becomes clear that these characteristics will be very difficult to approach with traditional selection methods. Although results from several studies in pigs indicate that quality characteristics are generally moderate in heritability, in ruminants and especially in sheep there is rather limited information on meat quality genetics. Hence, this section will concentrate on results from ruminants with some references to pigs. In addition, it is important to mention that nearly all the sheep meat quality results which will be reviewed below are breed comparisons and few are within-breed studies.

1.4.4.1   Colour

Meat colour is a function of myoglobin content and light-scattering properties of the muscle. Studies in beef (Dikerman, 1994) report that myoglobin content is moderately to highly heritable. The extent of light scattering and hence the lightness or darkness of meat, is dependent upon the myofibrillar volume. While no reports of heritability for the myoglobin content or light-scattering properties of muscle were found in lamb, Kuchtik *et al.* (1996) reported significant breed differences in reflectance values of joints. It is therefore likely that there is genetic variance in these traits and that they would respond to selection. Legrand *et al.* (1995) suggested that breed differences may occur with regard to fat quality, especially fat colour. They showed that ram lambs sired by Texels had a greater acceptability of colour score for the subcutaneous fat when compared with lambs sired by Charollais. Crouse *et al.* (1981) also reported breed effects on fat colour, with Suffolk sired lambs having yellower fat than lambs sired by Rambouillets.

1.4.4.2   Texture

As far as tenderness is concerned, Young *et al.* (1993) used a trained taste panel to examine the differences in consumer assessed tenderness of the loins of terminal progeny from Merino dams put to six sire breeds. They showed that the pure Merinos had a higher overall tenderness score than Oxford Down, Suffolk, Poll Dorset and Texel cross lambs; there were no differences between lambs sired by Border Leicesters and any of the other breeds.

In a different study, Sañudo *et al.* (1998) reported that the meat of Spanish lambs was significantly (p<0.05) less tender than meat of British lambs. In this study, there was no substantial evidence from the results to suggest that weight or age, significantly affected tenderness. On the other hand, British lamb is generally considered to be tender and less tough than other meats (Wood and Fisher, 1990; Sañudo *et al.*, 1998). This is probably due to an adequate fatness and chilling rate, which are important factors in determining eating quality. Thus, the higher tenderness of British lamb, in comparison to Spanish lamb, could be related to its higher fatness level and post-slaughter processing *i.e.* adequate protection from cold shortening and to differences in ageing time (Jaime *et al.*, 1993).

1.4.4.3   Juiciness

According to the previous study of Sañudo *et al.* (1998), juiciness was higher for Spanish lamb than for British lamb. These differences were statistically significant (p< 0.01) in the Spanish panel (17.8%) but less so in the British panel (p< 0.05), which judged Spanish meat to be only 3% more juicy than the British meat. These results agree with other findings that show that breed or weight effects are not important in juiciness, and that juiciness is slightly higher in younger lambs. However, no explanation is given for these findings. Additionally, previous findings have shown that older animals have meat with lower water-holding capacity (WHC) than younger animals (Sañudo and Sierra, 1982; Hawkins *et al.*, 1985).

1.4.4.4   Flavour

The uniqueness of lamb flavour has been investigated, and numerous reviews are available on the subject. Jacobson and Koehler (1963) examined volatile compounds from roasting lamb from three breeds (South-down, Hampshire, and Columbia). They found carbonyl compounds contributed to aroma, but no differences were detected between breeds for these compounds or other volatiles. Cramer *et al.* (1970a) compared three breeds (Rambouillet, Targhee, and Columbia) for mutton flavour intensity. Mutton flavour intensified as the fineness of the wool increased with breeds. In a second study (Cramer *et al.*, 1970b), five breeds (Romney, Hampshire, Columbia, Rambouillet, and Merino) were compared for intensity of mutton flavour. Mutton flavour intensity was similar between the breeds, but unsaturated fatty acid content was higher in the finer-wool breeds. Several other studies comparing breeds or sires (Fox *et al.* 1962, 1964; Dransfield *et al.*, 1979; Mendenhall and Ercanbrack, 1979; Crouse *et al.*, 1981) have been conducted, but differences in lamb flavour due to breed or sire were not observed. In a comparison of sire breeds (Dorper vs Suffolk), Duckett *et al.* (1999) reported fatty acid compositional differences between sire breeds and a greater flavour preference for Dorper-sired lamb. Elmore *et al.* (2000) reported higher levels of Maillard-derived volatiles and certain PUFA in intramuscular fat from Soays compared to Suffolks when fed various oils. Sañudo *et al.* (2000) reported that finishing system was more important than breed in determining fatty acid composition and flavour.

1.4.4.5   Intramuscular fat

While the contribution of intramuscular fat to eating quality of sheep meat is still not quantified, it would be wise to heed the pig-breeding experience. In earlier studies, intramuscular fat content in pig meat was not identified as a major contributor to eating quality and little emphasis was placed on changes that occurred in selection programmes which decreased back fat in pigs. It is now generally agreed that the decline in eating quality in the modern-day pig is associated with the decline in intramuscular fat content (Wood and Cameron, 1994). A number of studies have indicated variable and often low correlations between different fat depots in lambs (Olson *et al.*, 1976) and other species (*e.g.* pigs, Duniec, 1961; cattle, Renand, 1995), which suggests that a decrease in dissectible fat depots may be possible without concomitant changes in the intramuscular depot.

1.4.4.6   Fatty acid profile

Cameron *et al.* (1994) examined the changes that occurred in the lipid content of the adipose tissue and fatty acid profiles of subcutaneous fat from lines of Texel-Oxford and Scottish Blackface rams that had been divergently selected for carcass lean content. They showed that, although back-fat depth responded to selection, there were no significant changes in the individual fatty acid concentrations of subcutaneous fat or in the proportion of unsaturated fatty acids. They did, however, show that there were breed differences for the concentration of myristic acid (C14:0), with Scottish Blackface rams having higher concentrations than Texel-Oxford rams.

Furthermore, Cameron *et al.* (1994) showed that across breeds and lines the concentration of unsaturated fat in the subcutaneous depot was positively correlated with subcutaneous fat depth. In the Texel-Oxford line, selection for reduction in back-fat thickness lowered the lipid content within the adipose tissue from the subcutaneous depot. In contrast, there was no effect within the Scottish Blackface breed (Cameron *et al.*, 1994).

Webb and Casey (1995) showed that there were genetic differences in the fatty acid composition of subcutaneous adipose tissue. After correction for differences in maturity, the concentrations of palmitic (C16:0), palmitoleic (C16: 1) and stearic (C18:0) acids differed between South Australia mutton and Merinos and Dorpers. Breed influenced the proportions of myristic (C14:0), heptadecenoic (C17: 1) and oleic (C18:1) acids; however, when compared at equivalent levels of fatness, the breed differences in the proportion of C17:1 and C18:1 in the subcutaneous tissue were negligible. These results indicate that selection could change lipid contents within the fat depots.

1.4.5 Molecular genetics and meat quality

Considering the genetic improvement of meat quality it becomes clear, from the aforementioned, that these characteristics can in principle be improved through traditional selection methods, because there is evidence of genetic variation, although quality characteristics are generally moderate in heritability. However, identification of QTL influencing meat quality traits and their utilization via marker-assisted selection (MAS) has the potential to improve the rate at which genetic progress can be made.

Most of the progress in the area of molecular genetics of meat quality, to date, has been associated with major gene effects. A well-known example is the halothane gene.

Commercially, the halothane gene is of interest because it results in increased carcass lean content (Aalhus *et al.*, 1991; Pommier *et al.*, 1992). However, halothane reactors (*nn*) , in comparison with negative animals (*NN*) , are more stress susceptible and produce poor quality meat, particularly in terms of a higher incidence of pale, soft, and exudative (PSE) meat (Simpson and Webb, 1989; Sather *et al.*, 1991b,c; Jones *et al.*, 1994). The carcass composition and meat quality of carriers relative to negative animals is less clear, with reports in the literature that the gene may be recessive (Webb *et al.*, 1982) or additive (Jensen and Barton-Gade, 1985; Simpson and Webb, 1989). A possible explanation for differences in gene expression with respect to carcass composition and meat quality was proposed by Sather *et al.* (1991a,b), who observed a slaughter weight halothane genotype interaction for carcass lean content and meat quality traits. Another example of a major gene is the (*RN*) gene, affecting a measure of cured-cooked ham processing yield, the so called Rendement Technologique Napole (RTN) (Naveau *et al.*, 1985), which was first suggested by Naveau (1986), and then confirmed by Le Roy *et al.* (1990). The *RN* gene exists as two alleles; one is recessive (*rn+*, normal RTN) and one is dominant (*RN*, low RTN). The *RN* allele dramatically increases the *in vivo* muscle glycogen content (glycolytic potential; GP). The *RN* gene has been mapped on pig chromosome 15 (Milan *et al.*, 1995, 1996; Mariani *et al.*, 1996), and the mechanisms underlying its effects on muscle traits are unknown. In most studies reported so far, the influence of the *RN* allele on muscle and meat quality traits has been studied using pigs classified as carriers (*RN-/rn+* or *RN-/RN-*) or noncarriers (*rn+/rn+*) on the basis of high or low GP (Estrade *et al.*, 1993a, b; Lundström *et al.*, 1996; Enfält *et al.*, 1997a; Lundström *et al.*, 1998) or low or normal RTN values (Monin *et al.*, 1992).

The halothane and the *RN* gene are both examples where a major gene effect was identified, mapped and subsequently the gene was identified. In addition, both the candidate gene and genome scan approaches have been widely used to search for meat quality associated genes in pigs.

In sheep, some major genetic effects have been identified and reported. The most remarkable is that associated with the callipyge gene in sheep. This gene was first detected due to the highly significant effect it has on muscular hypertrophy; some individual muscles can be more than 40% heavier in older lambs expressing this trait compared to controls (Duckett *et al.*, 2000). The mode of inheritance is unique in that only animals that inherit the callipyge allele from their sire and the normal allele from their dam display the trait (*i.e.* any animal inheriting the callipyge allele from its dam does not display the trait) (Cockett *et al.*, 1996). The effect on meat quality is equally marked, animals expressing the callipyge trait have remarkably tough meat as assessed through shear-force measures or sensory panels, and the meat is associated with high calpastatin levels (Freking *et al.*, 1999; Duckett *et al.*, 2000). The effect of the callipyge locus on tenderness is perhaps the largest genetic effect known on this trait and precludes the sale of fresh meat from these animals without substantial ameliorative treatment.

Although there are no well-documented genome scans for meat quality associated QTL reported for sheep, there have been several genome scans performed in beef cattle. A well documented study in the literature is by Keele *et al.* (1999), where a family was derived from a *Bos Taurus x Bos indicus* (Hereford x Brahman) sire. Hence, this study focused on genetic effects differentiating these two sub-species. A major QTL was identified for which the Brahman derived allele decreased meat tenderness substantially. However, this QTL showed significant interaction with unknown environmental effects, such that it was observed in only one of four slaughter groups. Most studies using beef-type cattle evaluate growth, body composition and/or meat quality traits and several QTL have been detected for each category of traits (*i.e. fat yield (BTA 26)* - Stone *et al.* 1999; *hot carcass weight (BTA 6), marbling score (BTA 17 and 27)* - Casas *et al.* 2000; *marbling score (BTA 8), fat depth (BTA 8), Warner-Bratzler shear force (BTA 9)* - Casas *et al.* 2001; *live weight (BTA 17), marbling score (BAT 2)* MacNeil and Grosz 2002; *growth (BTA 2 and 6), yearling weight (BTA 5), hot carcass weight (BTA 23)*, Kim *et al.* 2003). Furthermore, in beef cattle there is another project called "RoBoGen" which was set up in 1998 at Roslin Institute with the aim of localising the genes affecting a wide range of important traits, especially those that would be difficult to measure on farms or in traditional selection programmes. The herd is a cross between breeds of cattle that differ greatly in their appearance and use; the most widely used dairy breed, the Holstein, and a beef breed, the Charolais. A number of QTL affecting meat quality have been identified, however the results remain confidential. These QTL will potentially benefit genetic selection programmes, especially carcass composition and meat quality traits, which are difficult and expensive to measure.

In summary, the use of molecular markers offers a great potential to improve efficiency of animal breeding. They also offer the potential for the identification of genes making significant contributions to variation. It is interesting to note that even in dairy cattle, where there has been intense selection for yield and significant selection for aspects of milk quality, there remain major genetic effects segregating (Haley, 2001). This suggests there are more effects to be found in under-explored species such as sheep, pigs, and in populations of other species that have not yet been investigated. Therefore, the challenge is to understand the biological processes that are involved in the development of economically important traits, such as meat quality, and thus, to increase the efficiency of livestock production and benefit consumers.

1.5   Thesis overview and objectives

This General Introduction has highlighted that there is a little knowledge on both quantitative genetics and identification of QTL for meat quality in sheep. The decline in lamb consumption together with the increased demand for improved meat quality by consumers over recent years has changed the selection objectives in sheep meat breeding. Therefore, sheep breeders need to address product quality traits. On the other hand, meat quality traits pose particular problems for improvement, as measurement is generally restricted to the slaughtered animal. Thus, new measurement technology, such as CT, offers the potential of more accurate measurement of carcass traits. In addition, CT traits could be used as predictor traits because direct measurement of meat quality is difficult, if not impossible, to measure in the live animal, and very expensive to measure completely in samples from the carcass of relatives of selection candidates. However, before this can be advocated it is necessary to explore the predictive relationships between CT measurements and meat quality attributes.

Hence, this thesis has been designed to target the following objectives:

> *Evaluate the utility of CT as a selection tool* (**Chapter 2**). Although some studies have demonstrated that CT scanning together with ultrasonic scanning can be used in breeding programmes to improve carcass traits, it is of interest to evaluate the impact of a CT index on carcass quality, when used as a part of a breeding programme.

> *Investigate the inheritance of carcass and meat quality in Scottish Blackface sheep (Chapters 3, 5 and 6).* Selection experiments have documented changes in carcass composition by selecting on live animal measures, but there is lack of information on the inheritance of traits assessed by CT scanning. In addition, there are no available genetic parameters for most of the meat quality traits, *e.g.* fatty acids and eating quality traits, in sheep.

28

➢ *Investigate relationships between carcass composition traits, as measured by CT, and meat quality traits (**Chapters 3, 5 and 6**).* As mentioned above, meat quality traits are interrelated, thus, it is important that those relationships are understood. In addition, relationships between CT assessed traits and meat quality have not been investigated yet and, hence, this might offer opportunities for predicting meat quality by CT traits.

➢ *Identify QTL for carcass composition and meat quality traits (**Chapters 4 and 5**).* Genetic improvement of meat quality is difficult to approach with traditional selection methods due to the fact that muscle quality is difficult and expensive to measure in the live animal. Hence, molecular genetics can assist in identifying genetic markers associated with meat quality. Although identification of genetic markers has progressed rapidly in farm animals, in sheep populations there has been little activity for traits such as fatty acid composition and eating quality. Thus, there is a need for molecular information for meat quality.

➢ *Explore selection index outcomes using the identified QTL (**Chapter 7**).* As a consequence of the aforementioned objectives, the final aim of this thesis is to explore possibilities for using the CT predictors of meat quality, and the identified QTL, in a selection index to improve various aspects of meat quality, *i.e.* improve flavour, healthiness and shelf-life of meat.

Chapter Two
Selection for carcass quality in hill sheep measured by X-ray computer tomography

2.1 Introduction

Genetic improvement of hill sheep is an important objective within the UK sheep industry because of the importance of this sector in terms of maintaining the rural environment and its contribution of 34% of the genes to the slaughter generation (Meat and Livestock Commission, 1998). A particular issue, in hill sheep, is that they have been naturally selected primarily for their ability to survive and produce lambs in harsh environments. This is thought to have tended to favour the development of smaller and hardier breeds, which produce carcasses of small size and of moderate conformation.

Several options are available for assessing carcass composition *in vivo*. Ultrasonic scanning is widely used in commercial breeding programmes as it is a measurement that can be readily used on large numbers of animals (McEwan *et al.*, 1989; Cameron and Bracken, 1992; Fennessy *et al.*, 1993; Afonso and Thompson, 1996; Simm *et al.*, 2002). Computer tomography (CT) scanning has been demonstrated to improve prediction of carcass traits and genetic improvement over and above ultrasonic scanning (Jopson, *et al.*, 1995; Young *et al.*, 1996, 1999), albeit at a considerably higher cost. If suitable breeding structures are in place, CT has been shown to be cost effective (Simm *et al.*, 1987; Allen, 1990; Jopson *et al.*, 1997). Jones *et al.* (2002) showed that CT scanning also provides good *in vivo* measures of muscularity. Muscularity measures are of interest as they provide objective description of the shape of a carcass independent of fatness. In the past, muscularity assessment could only be obtained either directly through carcass dissection after slaughter or by using ultrasound on live animals, as an indirect assessment of carcass tissue.

A long-term experiment at ADAS Redesdale was initiated in 1992, to improve carcass quality (weight and conformation) in hill sheep through genetic selection. Whereas good progress was made in increasing carcass weights (Roden *et al.*, 2003), progress towards improving carcass conformation was hampered by inadequate live-animal measures of this trait. Thus, in 1997, a study was initiated in this population to explore the possibility of using CT measures to improve carcass quality. An index of CT traits was derived to predict carcass conformation and composition, and selection on this index commenced. The aim of this **Chapter** is to evaluate the impact of this CT index on carcass quality, when used as part of a breed improvement programme.

2.2 Materials and Methods

2.2.1 Flock establishment and experimental design

The thesis is based on two separate animal materials; the first one is investigated in this **Chapter**, and the second one in the remaining **Chapters (3, 4, 5, 6, and 7)**. The design in this **Chapter** was based in an open nucleus flock of 60 Scottish Blackface donor ewes that was established in 1992 from a base population of approximately 1600 ewes maintained, under hill conditions, at ADAS Redesdale in Northumberland, England. The selection history of these base flocks was one of objective visual inspection to eliminate faults and to improve body size, conformation and breed characteristics (Roden *et al.*, 2003). The selection criterion used from 1992 until 1998 was a desired gains index designed to increase live weight at ultrasonic scanning and ultrasonic muscle width and depth, without changing ultrasonic fat depth, with the aim that this should also improve the characteristics of the carcass, *i.e.* conformation score and fat class (Roden *et al.*, 2003).

The flock was maintained, and selection performed, using a multiple ovulation and embryo transfer (MOET) procedure. In mid November each year all nucleus donor and recipient ewes were initially transferred from open hill grazing to in-bye fields. They were then housed for a short period immediately before and during the mating period (a total of approx. 2 weeks). This was done to facilitate the induction of superovulation, semen collection, AI, embryo collection and embryo transfer. Details of the treatment protocols for donor and recipient ewes and the methods for collecting and transferring embryos have been provided elsewhere (Bari *et al.,* 1999 and 2000). The MOET design of the breeding project allows maternal environmental effects on lamb performance to be separated from direct genetic effects.

From 1998 onwards, each year the flock comprised 60 donor ewes mated with six rams chosen on the basis of their carcass characteristics (described below), and *ca.* 300 recipient ewes (details in Roden *et al.*, 2003). Following the MOET procedure, these 60 ewes were then naturally mated to 2 further rams, also chosen on the basis of their carcass characteristics. Additionally, each year *ca.* 40 control ewes were naturally mated to 4 rams from the same sub-population to provide a control line. During the course of this study, donor ewes were selected from amongst those born to selected rams on the basis of their carcass characteristics and adherence to industry-accepted breed-type characteristics. Recipient ewes were generally unrelated sheep, born outside of the nucleus.

2.2.2 Husbandry and phenotypic measurements

Male lambs were reared entire and all lambs were weaned at 20 weeks of age. Each year, from 1993 to 2003, all lambs were ultrasound scanned on the same day; the average age varied between years from 20 to 29 weeks. CT measurements were taken on lambs at 26 weeks of age, from 1998 to 2003, with the exception of 2001 when the Foot and Mouth Disease (FMD) epidemic and the subsequent livestock movement restrictions meant that CT measurements could not be taken on 2001-born lambs. In total the dataset comprised 2562 lambs recorded at birth for eleven years (1993-2003), these lambs being the progeny of 123 sires, 348 donors and 1229 recipients. Additional pedigree information was available, and the complete pedigree comprised 3799 animals.

Ultrasonic scanning was performed at the level of the 3rd lumbar vertebra using a Dynamic Imaging real-time ultrasound scanner, with a 7.5 MHz 56-mm probe. Live weight at scanning (SWT) and three individual subcutaneous fat depth measurements were taken over the *longissimus dorsi* muscle, at a constant age, moving laterally between the vertical and transverse processes of the vertebra, from which the average fat depth was calculated (UFD). A single muscle depth (UMD) and a single muscle width (UMW) measurement were taken at the deepest and widest point on the muscle respectively. Ultrasonic scanning data were available on 2018 of the lambs, representing 123 sires, 290 donors and 1071 recipients.

Carcass composition was assessed by Computed Tomography (CT) techniques on 1266 lambs, representing 49 sires, 189 donors and 711 recipients. CT measurements provided areas of fat, bone and muscle at the 8th thoracic vertebra (TV8), at the 5th lumbar vertebra (LV5) and the ischium (ISC). The traits were CT assessed area of tissue at each anatomical site.

Carcass assessments were taken on male lambs not required for further breeding purposes. Lambs were selected for slaughter at a target of fat class 3 (EUROP, 1995), by the same person on all occasions. Measurements included slaughter weight and carcass weight, conformation score (EUROP, 1995) and fat class, and total price received for each animal was recorded. Carcass data were available for 1148 lambs, representing 79 sires, 124 donors and 799 recipients.

2.2.3 CT selection criteria

A 'desired gains' index was derived from available unpublished data from a CT calibration trial on Blackface lambs, unpublished data from the SAC/Roslin Hill Sheep Project and data collected on these sheep in 1998, with the aim of simultaneously improving both carcass composition and carcass conformation. The derived index, combining live weight (taken on the farm, prior to CT scanning), subjective conformation, and the fat, muscle and bone areas in the ischium, 5th lumbar vertebra and 8th thoracic vertebra, was:

CT index = -216.6(Live weight) + 272.2 (Subjective Conf.) –0.00596 (ISC fat) + 0.1814(ISC muscle) + 0.2838 (ISC bone) + 0.0263 (LV5 fat) + 0.169 (LV5 muscle) + 3.156 (LV5 bone) + 0.128 (TV8 fat) + 0.139 (TV8 muscle) + 0.273 (TV8 bone)

Subjective conformation score was the average of assessments on the live animal at ultrasound scanning at the shoulder, leg and loin, scored from 1 (very lean) to 5 (very fat). This index was derived with the following relative 'economic weightings': conformation = 2, fat = -1, muscle = 2. These weightings were chosen because, given the available data, they resulted in predicted gains in carcass characteristics that were close to perceived optimal responses, *i.e.* improved conformation and decreased fatness with minimal changes in live weight. Additionally, they reflected the economic weightings of fat = -1 and muscle = 2 advocated by Simm and Dingwall (1989) and now widely used in UK terminal sire selection indices.

The CT index was implemented in 1999, and the first cohort of lambs born as a result of selection decisions made using the CT index were born and evaluated in 2000. Each year, BLUP estimated breeding values (EBVs) were calculated for the CT index for all animals, and sires were selected on the basis of the Index EBV along with adherence to industry-accepted breed-type characteristics. As described above, animals were not CT scanned in 2001. To enable selection to continue in 2001, lambs were ranked on an index comprising their parental average CT index EBV and their muscle width EBV. Based on data available at the time and using selection index methodology, this index had a relative efficiency of 0.75 compared to the CT index based on data from the selection candidates themselves.

2.2.4 Statistical analysis

2.2.4.1 Data summary and trait definition

Data analyses were performed using all the available data, however responses to selection are only presented for the years in which CT measurements were taken. Data were initially analysed using multiple regression techniques in order to identify significant fixed effects, covariates and two-way interactions, using Genstat (2003). The fixed effects subsequently included in the analysis for slaughter and carcass traits were: year of birth (11 classes), lamb sex (male or female), type of rearing (singles and twins), and flock (2 classes). Recipient age (7 levels) was used, as a fixed effect, only in the analysis of live animal measurements. Day of birth and weaning age were fitted as covariates for ultrasound scanning traits and CT traits. Analyses of traits measured post slaughter were performed with unadjusted data, or data adjusted to either a constant fat classification score or to a constant age at slaughter, fitting these effects as covariates. Adjustment to constant fat classification score were made by linear regression and adjustment to a constant age was by linear plus quadratic regressions on age at slaughter. Fixed effects and covariates found to be significant in the multiple linear regression analysis were included in the variance component analysis of each trait. Day of birth was not found to be significant for the post slaughter traits.

To analyse post slaughter fat classification (EUROP, 1995), *i.e.* the amount of fat on the outside of the carcass visible to an assessor, a new trait (fat class) was constructed as follows: fat classification 5 (very high)→5, 4H→4, 4L→3.5, 3H→3, 3L→2.5, 2 (low)→2 and 1 (very low)→1. Also, conformation score values were recoded as E (Excellent)→5, U (Very Good)→4, R (Good)→3, O (Medium)→2 and P (Weak)→1. Thus, higher scores indicate better conformation and greater fatness. Conformation describes carcass shape in terms of convex/concave profiles and indicates the amount of flesh (muscle + fat) in relation to the length of bones. Total price was the price paid by the abattoir for the carcass.

2.2.4.2 Genetic parameters

Restricted maximum likelihood methods were used to estimate variance components using an animal model, fitting the complete pedigree structure (3799 animals), using *ASReml* program (Gilmour *et al.*, 2004). All traits, except the CT traits, were initially analysed in three univariate models. Model 1 fitted animal as the only random effect

(Model 1): $y = Xb + Za + e$. In model 2, the genetic component of the maternal (i.e. recipient) effect was included by fitting recipient as a second random effect, uncorrelated with the animal genetic effect but accounting for pedigree relationships between recipient dams: $y = Xb + Za + Wm + e$. In model 3, the effect of common environment was included by fitting litter as a third uncorrelated random effect: $y = Xb + Za + Wm + Sc + e$. In these models, y is the vector of observations on the specific trait of the animal; b is the vector of fixed effects, described before; a is the vector of additive random animal (genetic) effects, m is the vector of random recipient effects, c is the vector of random common environmental effects (litters) and e is the vector of random residual effects. X, Z, W and S are the incidence matrices relating records to fixed, animal, recipient and common environmental effects respectively. It was assumed that the expectations (E) of the variables were: $E(y) = Xb$; $E(a) = 0$; $E(e) = 0$; $E(c) = 0$ and that common environmental effects and residuals effects are independently distributed with means of zero and variances σ_c^2 and σ_e^2 respectively. Therefore, $var(a) = A\sigma_a^2$; $var(m) = A\sigma_m^2$; $var(c) = I\sigma_c^2$; $var(e) = I\sigma_e^2$, where A is the numerator relationship matrix of animals in the model; I is the identity matrix; σ_a^2 is the additive genetic variance for direct effects (animal), σ_m^2 is the additive genetic variance for recipient effects, σ_c^2 is the variance due to common environmental effects and σ_e^2 is the residual error variance. Due to the fact that models were nested, the significance of the second and third random effects were tested using log-likelihood ratio tests, to determine the most suitable model for each trait in univariate analyses. Additionally, the univariate analyses were used to obtain starting values to use in subsequent bivariate analyses.

Each combination of traits was then analysed with a bivariate model using the most appropriate model for each trait. Approximate standard errors for heritabilities and correlations were constructed by ASReml package (Gilmour *et al.*, 2004) from approximations to variances of ratios and products.

2.2.4.3 Genetic trends and selection differentials

Mean values for the selection line (**S**) and the control line (**C**) were calculated by two methods: (i) calculating the average CT index EBVs for **S** and **C** line animals each year when fitting model 1 (animal as the only random effect), and (ii) by fitting line x year as an additional fixed effect in the REML analysis. Genetic trends for the direct additive genetic values were then calculated by regressing the line difference (**S-C**) for each year on year of birth. The impact of selection on other traits was estimated from the difference between the selection and control line fixed effects, averaged across 2002 and 2003 birth years. The average was taken across two years, as the small size of the control line meant that comparisons made from any one year were imprecise. The standard errors of these estimated differences were constructed from the variance/covariance matrix of fixed effects.

Selection differentials were estimated to investigate the extent to which selection was truly performed using the CT index. The actual (effective) selection differential (EFSD) per year was calculated as the weighted average CT index EBV of sires against the mean EBV of the cohort from which they were selected, the weight given to each sire being his proportional contribution to the individuals that were measured in the next generation. The maximum potential selection differential (MPSD) was estimated by calculating the difference between the mean EBV of the six top ranking sires and the mean EBV of the cohort from which they would have been selected, using the CT index as the sole ranking criterion and the same selection intensity as was achieved in practice. Comparing the selection differential achieved and the maximum potential provides an estimate of the proportion of the possible selection that was applied to the selection criterion in the selection line. The actual expected and maximum potential selection differentials were all estimated within year (Mrode *et al.*, 1990).

2.3   Results

2.3.1 Summary statistics

Mean values for the CT, ultrasonic and carcass measurements are shown in Table 2.1 along with the phenotypic standard deviations for each trait. Backfat depth in Blackface lambs was an extremely variable trait with a coefficient of variation of 0.61 compared to 0.11 for muscle depth. Also, the three fat areas (ISC, LV5, TV8) assessed by CT were moderately variable traits with coefficient of variations of 0.28, 0.54 and 0.38 respectively.

Table 2.1 *Mean values and phenotypic standard deviations for CT, ultrasound and carcass traits[t]*

Trait	n	Mean	Phenotypic s.d.[t]
CT index (units)	1074	4621	475
CT traits			
Live weight (kg)	1266	32.7	3.27
Fat area ISC (mm^2)	1266	3504	994
Muscle area ISC (mm^2)	1266	20560	1613
Bone area ISC (mm^2)	1266	2887	269
Fat area LV5 (mm^2)	1266	1326	713
Muscle area LV5 (mm^2)	1266	7270	772
Bone area LV5 (mm^2)	1266	859	106
Fat area TV8 (mm^2)	1266	3511	1330
Muscle area TV8 (mm^2)	1266	9451	1095
Bone area TV8 (mm^2)	1266	3303	425
Ultrasound traits			
Muscle width (UMW) (mm)	2018	41.5	3.32
Muscle depth (UMD) (mm)	2018	20.8	2.21
Fat depth (UFD) (mm)	2018	1.70	1.04
Scan weight (SWT) (kg)	2554	32.7	3.50
Post slaughter traits[‡]			
Carcass weight (kg)	1148	18.3	1.98
Slaughter weight (kg)	1148	42.8	4.22
Conformation score (units)	1148	2.97	0.49
Fat class (units)	1148	2.75	0.50
Total Price (£)	485	30.9	4.36

[t] Phenotypic standard deviation obtained from REML analyses.
[‡] Data not corrected to a predefined endpoint.

2.3.2 Line differences and responses to selection

Mean values for the CT index for the **S** and **C** lines, from REML analyses fitting the line x year interaction, are in Figure 2.1 and responses to selection on the CT index, as estimated from the difference between the **S** and **C** lines, are in Figure 2.2. The estimated genetic trend was 52 (s.e. 38) units/year for the CT index, representing an annual genetic progress of 1.2% relative to the base year or 0.11 phenotypic standard deviation units per year. The genetic trend for the CT index estimated from the line mean EBV was 40 (s.e. 24) units/year. Although it appears from Figure 2.1 that the control line may also have drifted upwards with time, these year by line means do not disentangle genetic and environmental effects.

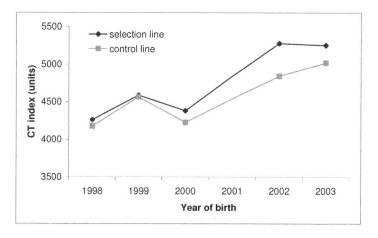

Figure 2.1 Line means for CT index for selection (S) and control (C) lines

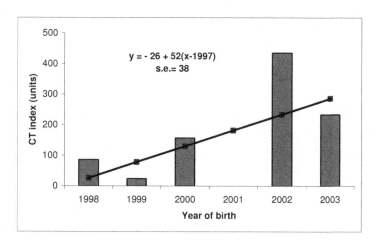

Figure 2.2 Direct genetic trends for CT index from 1998 to 2003, with the straight line showing the regression of line difference on year

Line differences (**S-C**) for the CT and ultrasonic assessed traits are shown in Table 2.2, and in 2.3 for carcass traits, with significant (*P*< 0.05) **S-C** line differences shown in bold. The **S** line lambs had greater muscle and bone areas at the three sites (ISC, LV5, and TV8) and greater ultrasonic muscle width and depth. Furthermore, the **S** line lambs had higher conformation score and lower fat class score, both of which were in the desired direction. Non-significant trends were also seen for fat areas at the three sites (ISC, LV5, and TV8), with the **S** line having less fat than the **C** line. Line differences for carcass traits adjusted to a constant fat classification score and to a constant age were similar to those described above.

Table 2.2 *Achieved correlated responses for traits measured in the live animal, assessed using data collected in 2002 and 2003 (significant P< 0.05 line differences are show in bold)*

Trait	Selection line (S)	s.e. (S)	Control line (C)	s.e. (C)	Line differences (S-C)	s.e.d
Live weight (kg)	28.66	0.24	27.54	0.44	**1.12**	0.49
Fat area ISC (mm^2)	2933	72.0	3010	135	-76.8	147
Muscle area ISC (mm^2)	20040	116	18888	217	**1153**	237
Bone area ISC (mm^2)	2666	19.4	2569	36.3	**97.3**	39.7
Fat area LV5 (mm^2)	712	52.3	738	97.9	-26.3	106
Muscle area LV5 (mm^2)	6929	55.4	6393	103	**537**	113
Bone area LV5 (mm^2)	838	7.69	807	14.4	**30.8**	15.7
Fat area TV8 (mm^2)	2497	97.0	2640	181	-142	198.3
Muscle area TV8 (mm^2)	9687	79.1	8919	148	**768**	161.7
Bone area TV8 (mm^2)	2888	31.2	2714	58.4	**175**	63.8
Muscle width (UMW) (mm)	40.33	0.22	38.50	0.44	**1.83**	0.48
Muscle depth (UMD) (mm)	19.74	0.14	18.87	0.29	**0.87**	0.32
Fat depth (UFD) (mm)	1.12	0.07	1.06	0.13	0.06	0.15
Scan weight (SWT) (kg)	25.49	0.23	24.17	0.46	**1.32**	0.50

Table 2.3 *Achieved correlated response for carcass and slaughter traits[t], assessed using data collected in 2002 and 2003 (significant P< 0.05 line differences are show in bold)*

Trait	Selection line (S)	s.e (S)	Control line (C)	s.e. (C)	Line differences (S-C)	s.e.d.
Carcass weight (kg)	16.3	0.23	16.8	0.37	-0.58	0.41
Slaughter weight (kg)	37.1	0.48	38.1	0.80	-1.00	0.87
Conformation score (units)	2.96	0.06	2.74	0.09	**0.22**	0.10
Fat class (units)	2.49	0.06	2.78	0.10	**-0.28**	0.10
Total Price (£)	28.1	0.60	28.3	1.02	-0.29	1.12

[t] Data not corrected to a predefined endpoint.

2.3.3 Genetic parameters

Univariate heritability estimates for all live animal measurements are presented in Table 2.4. Maternal genetic effects were significant, and therefore fitted, for some traits. The model containing common environmental effects (litter) was never significantly better than that with maternal genetic effects. Heritabilities for traits describing carcass composition (*i.e.* ultrasonic and CT measures) were all moderate, and the maternal genetic components for these traits were small or not significant. In contrast, for live weight at scanning, the maternal contribution and the contribution of the genes of the lamb are very similar (0.16 and 0.17, respectively). The CT index was highly heritable (0.41), which is consistent with the observation that selection on the index was successful.

Table 2.4 *Univariate heritabilities (h^2), maternal effects (m^2) both with standard errors (s.e.) for traits[†] measured in the live animal*

Trait	h^2	s.e. (h^2)	h_m^2	s.e.(h_m^2)
CT index (units)	0.41	0.08	-	-
Live weight (kg)	0.19	0.06	0.14	0.04
Fat area ISC (mm^2)	0.34	0.07	-	-
Muscle area ISC (mm^2)	0.38	0.07	0.12	0.04
Bone area ISC (mm^2)	0.37	0.08	-	-
Fat area LV5 (mm^2)	0.30	0.07	-	-
Muscle area LV5 (mm^2)	0.38	0.07	0.11	0.04
Bone area LV5 (mm^2)	0.35	0.07		
Fat area TV8 (mm^2)	0.26	0.07	0.10	0.04
Muscle area TV8 (mm^2)	0.39	0.07	-	-
Bone area TV8 (mm^2)	0.29	0.07	-	-
Muscle width (UMW) (mm)	0.14	0.04	-	-
Muscle depth (UMD)(mm)	0.41	0.05	-	-
Fat depth (UFD)(mm)	0.30	0.05	0.06	0.03
Scan weight (SWT)(kg)	0.16	0.03	0.17	0.03

[†] Data not corrected to any endpoint.

Heritabilities for post slaughter traits are presented in Table 2.5. Maternal and litter effects were largely unimportant for these traits. The end-point adjustment (none, fat class or age) had no impact on the heritability values, however it did affect the standard errors of the heritabilities: they were lowest when a quadratic age correction was used and it is these values that are presented in Table 2.5. Almost all these traits were lowly heritable, except for carcass weight and slaughter age, which were both moderately heritable (0.21).

Table 2.5 *Univariate heritabilities (h^2), with standard errors (s.e.) for carcass and slaughter traits at a constant age*

Trait	h^2	s.e.
Carcass weight (kg)	0.21	0.06
Slaughter weight (kg)	0.15	0.05
Conformation score (units)	0.14	0.05
Fat class (units)	0.19	0.05
Slaughter age (days) [+]	0.21	0.05
Total Price (£)	0.19	0.06

[+] Not adjusted to a constant age.

The estimated phenotypic and genetic correlations between the CT index, ultrasonic and carcass traits are shown in Table 2.6, with mean heritability values calculated from the bivariate models. The heritability values presented in Tables 2.4 and 2.5 (using univariate models) and Table 2.6 (using the average of the bivariate heritability estimates) are very similar. The estimates of phenotypic correlations are robust, with no standard errors greater than 0.06. However, there are large standard errors for several genetic correlations. This is particularly the case where one or both traits have low heritabilities and it limits the interpretation of many of the correlations presented.

Table 2.6 *Heritabilities (using the average of the bivariate heritability estimates), phenotypic and genetic correlations (with standard errors) for CT, ultrasound and carcass traits[††]*

Trait	CT Index	SWT	UMW	UMD	UFD	Slaughter weight	Carcass weight	Conformation score	Fat class	Total price
CT Index	**0.41**	0.14	0.16	0.35	0.15	0.12	0.12	0.00	-0.07	0.04
	(0.07)									
SWT	0.32	**0.20**	0.34	0.34	0.33	0.52	0.41	0.05	0.07	0.02
	(0.16)	**(0.04)**								
UMW	0.18	0.24	**0.14**	0.36	0.20	0.18	0.17	0.05	0.05	0.01
	(0.18)	(0.16)	**(0.04)**							
UMD	0.50	0.23	0.30	**0.40**	0.32	0.16	0.17	0.15	0.12	-0.11
	(0.10)	(0.12)	(0.12)	**(0.05)**						
UFD	0.10	0.10	-0.05	0.17	**0.41**	0.16	0.20	0.10	0.20	-0.16
	(0.15)	(0.13)	(0.15)	(0.10)	**(0.08)**					
Slaughter weight	0.19	0.95	0.12	0.09	-0.02	**0.12**	0.84	0.09	0.18	0.61
	(0.22)	(0.07)	(0.21)	(0.15)	(0.17)	**(0.09)**				
Carcass weight	0.40	0.87	0.06	0.24	0.09	0.91	**0.17**	0.13	0.26	0.74
	(0.26)	(0.16)	(0.26)	(0.18)	(0.21)	(0.13)	**(0.09)**			
Conformation score	0.13	-0.17	-0.01	0.30	-0.08	0.11	0.01	**0.22**	0.21	0.06
	(0.28)	(0.24)	(0.26)	(0.19)	(0.22)	(0.45)	(0.37)	**(0.10)**		
Fat class	-0.49	0.16	-0.02	0.20	0.97	0.20	0.20	0.19	**0.25**	0.09
	(0.25)	(0.20)	(0.24)	(0.17)	(0.17)	(0.42)	(0.35)	(0.28)	**(0.13)**	
Total price	0.35	0.63	-0.37	0.11	0.13	0.63	0.90	-0.32	-0.22	**0.16**
	(0.34)	(0.32)	(0.36)	(0.26)	(0.30)	(0.23)	(0.34)	(0.51)	(0.48)	**(0.08)**

[†] Heritabilities on diagonal, phenotypic correlations above and genetic correlations below diagonal.
[‡] Data not corrected to a predefined endpoint.

The genetic correlations between the CT index and ultrasonic measurements were all positive and generally low to moderate. The estimated genetic correlations between the CT index and carcass traits were generally positive and low to moderate, except that with fat class which was negative.

The genetic correlations between ultrasonic measurements were positive and low to moderate with large standard errors, and in the case of backfat depth, not significantly different from zero. In addition, carcass traits were moderately positively correlated, with a very high genetic correlation for carcass and slaughter weight (0.91). Genetic correlations between ultrasonic and carcass traits were variable. Live weight at scanning was highly positively correlated with slaughter and carcass weight.

The estimates of total price with slaughter age and live weight at scanning, and those of slaughter and carcass weight with slaughter age are interesting. The genetic correlation between total price and carcass weight is positive and very strong and the genetic correlation between price/kg and carcass weight is, also, positive and strong (0.52, s.e. 0.14) (result not shown). Also, the genetic correlation between total price and slaughter age is positive and very strong (0.90, s.e. 0.20) (result not shown). Additionally, the estimated genetic correlations of total price with live weight at scanning and slaughter weight were positive and very strong (both 0.63). Slaughter age had a strongly positive genetic correlation with slaughter (0.44, s.e. 0.15) and carcass weight (0.71, s.e. 0.15).

2.3.4 Actual and maximum selection differentials
Table 2.7 shows the ratio of the selection differential achieved to the maximum potential selection differential for the CT index. This ratio provides an evaluation of how effective actual selection was relative to the intended selection. In the **S** line, the selection differential achieved for CT index across all years was proportionately 0.89 of the maximum potential. The selection differential achieved for CT index was proportionately 0.91 for 1999 and 2000, 0.87 for 2001, and 0.88 for 2002 and 2003, all relative to the maximum potential for that year. The value achieved in 2001, when indirect selection had to be practised, was only marginally lower than that achieved in other years.

Table 2.7 *Ratio of actual to maximum selection differentials achieved for selected sires for the CT index each year*

Year	Actual/Maximum
Overall mean	0.89
1999	0.91
2000	0.91
2001	0.87
2002	0.88
2003	0.88

2.4 Discussion

2.4.1 Inheritance of traits

The heritability estimates for CT traits, *i.e.* both the CT index and the CT tissue areas, on the live animal were moderate, indicating that CT provides a quick and reliable means of genetically changing carcass composition in sheep. In another study, on a different population of Blackface sheep at 24 weeks of age (**Chapter 3**), the heritability estimates of CT tissue areas traits were moderate to high, being very similar for bone and muscle areas (average $h^2 = 0.36$ and average $h^2 = 0.33$, respectively, but somewhat higher than the current study for the three fat areas (average $h^2 = 0.64$).

The ultrasonic measures of post-weaning body composition were moderately to highly heritable. Genetic parameters for ultrasonic measurements have previously been reported in this population. The estimate of heritability for fat depth ($h^2 = 0.44$) was lower than that reported by Roden *et al.* (2003) on a subset of this dataset, by Puntila *et al.* (2002) for Finnsheep ($h^2 = 0.39$), by Bishop *et al.* (1996) for Scottish Blackface sheep kept in an intensive environment ($h^2 = 0.39$), by Saatci *et al.* (1998) for Welsh Mountain sheep ($h^2 = 0.40$), by Thorsteinsson and Eythorsdottir (1998) for Icelandic sheep ($h^2 = 0.42$), and by Cameron and Bracken (1992) for a terminal sire breed ($h^2 = 0.35$). Our estimate for fat depth is almost the same with that summarized by Fogarty (1995) for dual-purpose sheep (a mean of 0.28 (no.= 30) for fat depth). The heritability estimate for muscle depth coincide with that reported by Puntila *et al.* (2002) for Finnsheep ($h^2 = 0.46$) and by Cameron and Bracken (1992) for a terminal sire breed ($h^2 = 0.46$). However, when considering other studies, the heritability of muscle depth in this study is high (Roden *et al.*, 2003 ($h^2 = 0.26$); Fogarty *et al.*, (2003) ($h^2 = 0.27$); Bishop *et al.*, 1996 ($h^2 = 0.25$) and 1998 ($h^2 = 0.20$); Saatci *et al.,* 1998 (h^2= 0.19)). Lauridsen (1998) found 25 literature estimates that the heritability of muscle depth ranged from 0.13 to 0.57 (average of 0.32) and of fat depth from 0.07 to 0.62 (average of 0.31). Usually fat depths have been reported to have medium to high heritabilities. These results suggest that the potential for genetic progress for body composition traits in hill sheep maintained under extensive conditions may be higher than previously thought, but they may also vary greatly between populations. The heritability of ultrasonic muscle width is considerably lower than that of muscle depth, and is in agreement with the study by Fogarty *et al.* (2003), where the heritability for muscle width in 19 months-old Merino rams was 0.15. This most probably reflects the fact that this parameter was more difficult to measure accurately than was muscle depth. Interestingly, heritabilities for fat and muscle measures from CT were almost the same with ultrasonic measures in this study, whereas in the study that will be presented in **Chapter 3**, CT heritabilities were higher than those from ultrasound, particularly for fat measurements.

The estimated heritability for live weight at ultrasound scanning was lower in this study than the estimates reported by Cameron and Bracken (1992) for a terminal sire breed (h^2= 0.20), Bishop (1993) for Scottish Blackface sheep (h^2=0.23) and Puntila *et al.* (2002) for Finnsheep (h^2= 0.44). However, a substantial maternal genetic component was also observed in this study.

Heritability estimates for carcass and slaughter traits were moderate. The heritability for carcass weight (h^2= 0.21) was in agreement with the study by Bennett *et al.,* (1991) (h^2= 0.22), which was done in crossbred lambs from Southdown sires and Romney ewes. Also, Henniñgsson and Malmfors (1995) reported an estimate of 0.17 for Swedish Pelt sheep and Conington *et al.* (2001) reported an estimate of 0.33 for Blackface sheep.

Heritability estimates for fat class and conformation score were low, as expected, given that they are subjectively scored. There are very few estimates on genetic parameters for visually assessed muscling on live animals, such as conformation score. Bradford and Spurlock (1972) reported a heritability estimate of 0.32 (s.e. 0.25) for live conformation, and Schrooten and Visscher (1987) obtained a relatively high heritability estimate for live muscling score (0.44, s.e. 0.09) in the Texel breed. van Heelsum *et al.* (1999) reported a heritability estimate of 0.20 (s.e. 0.04) for scores of gigot conformation in the Bluefaced Leicester and Puntila *et al.* (2002) reported an estimate of 0.27 for Finnsheep. Also, Conington *et al.* (2001) reported a heritability estimate of 0.09 for conformation and 0.17

for fat class score in Blackface sheep. The estimate for fat class score was in close agreement with our estimate. Heritabilities for live animal conformation score suggest that there is scope for genetic improvement in the breeds with low subcutaneous fat cover. It is known that conformation is strongly dependent on fatness and muscling (having moderate heritabilities) but scored in a subjective manner.

2.4.2 Relationships between traits

This dataset was large enough to allow estimation of genetic correlations between the CT index and ultrasonic and carcass measurements. The genetic correlations between the CT index and ultrasonic and carcass measurements were low to moderate, albeit with large standard errors. Genetic change in any of these attributes is feasible in principle, and correlated responses in other traits predictable from genetic theory.

The relationships between CT index and live weight at scanning and muscle depth were moderately strong, indicating that selection on the index should moderately increase both these traits. Also, CT index was strongly genetically correlated with carcass weight and fat class score, again indicating that selection on the index should increase carcass weight, but decrease fatness score.

A high genetic correlation for live weight at scanning with slaughter and carcass weight was found, indicating that selection for heavier lambs at weaning will increase the slaughter and carcass and slaughter weight. These results contradict the study made in Blackface sheep by Conington *et al.* (1998), where they found correlations of 0.18 and 0.07, respectively. Also, the very strong genetic correlation between ultrasonic fat depth and fat class indicates that selection for reduced fat depth could lower the fatness score, which is in the expected direction. In addition, carcass weight and slaughter weight were strongly genetically correlated, indicating that they are essentially expressions of the same trait in agreement with the study of Conington *et al.* (1998) for Blackface sheep. These correlations between live animal and slaughter measurements are likely to change with the production environment, hence, the mean time between weaning and slaughter and, critically, the between-animal variation in this time.

The relationships between total price and live weight at scanning, slaughter and carcass weight, were interesting. Obviously, the relationship between total price and carcass weight was positive and very strong, indicating that heavier lambs made more money. The estimates for total price and slaughter and carcass weight were in close agreement with those reported by Conington *et al.* (1998), where they found estimates of 0.92 and 0.89, respectively. However, this relationship is stronger than can simply be explained by the fact that bigger carcasses fetch higher prices because they are bigger, because price/kg was also genetically positively correlated with carcass weight. In other words, larger carcasses are favoured, financially. Also, the estimates for slaughter age with total price, slaughter and carcass weight were in agreement with the study by Conington *et al.*, (1998), where they report estimates of 0.84, 0.67 and 0.67, respectively. The relationship between slaughter age and total price indicates a trend in prices through the season that is counter to most selection goals, as it indicates that genetically slower growing animals will return more money to the producer, under the pricing structure encountered in this study.

2.4.3 Selection responses

Selection differentials provide a convenient description of the selection pressure that has been applied each year. Comparison of actual and maximum selection differentials indicate that selection was largely on the CT index, but that other criteria were also included in the selection decisions. Moreover, selection was not markedly hindered by the absence of CT measurements for one year. Selection on the index has been successful in increasing live weight and traits describing muscle dimensions, but reductions in fat dimensions, as assessed on the live animal were not significant. At slaughter this corresponded to a small increase in conformation score and a small decrease in fatness.

2.5 Conclusions

This study has confirmed that CT and ultrasonic measurements, in principle, provide an effective means of selecting for improved carcass composition. The ultrasonic and CT measured traits were moderately heritable, with the subjectively assessed traits being less heritable. Many of these traits were also affected by the maternal environment, as would be expected. The ultrasonic and CT traits were also very variable, which is the second requirement for making genetic progress.

Judging by the genetic correlations, selection on the CT index should achieve a moderate improvement in conformation, price and slaughter weight, and a decrease in fat class. These changes were in fact seen for fat class, conformation and live weight. Conversely, selection on live weight at scanning would be expected to improve carcass and slaughter weight, and total price received, but have slightly deleterious impacts on carcass conformation score.

Chapter Three
Genetic analyses of carcass composition, assessed by X-ray computer tomography, and meat quality traits

3.1 Introduction

Meat consumption has increased markedly in recent decades in the western world, with the exception being the consumption of lamb (Lewis *et al.*, 1993). At the same time as the increase in meat consumption, consumer demands on the quality of meat have risen. It is shown that consumption of meat is related to consumer satisfaction of meat, as outlined by Ward *et al.* (1995). Consequently, high meat quality stimulates consumption, which is encouraging for the meat industry.

Traits determining product quality are carcass composition and meat quality. A number of selection experiments have documented rapid changes in composition that are possible by selecting on live animal estimates of carcass weight or composition (*e.g.* Simm and Dingwall, 1989; Bishop, 1993; Jopson *et al.*, 1995). New measurement technology, such as computerised tomography (CT), offers the potential for more accurate measurement of carcass traits in the live animal (Sehested, 1984; Young *et al.*, 1987, 1996 and 1999) and consequently improved genetic gains, although this will need to be balanced against the increased costs. On the other hand, meat quality traits pose particular problems for improvement, as direct measurements require destruction of the animal. Meat quality is generally considered to be difficult, if not impossible, to measure in the live animal and is very expensive to measure completely in samples from the carcass (Clutter, 1995).

In addition to measurement issues, as already mentioned in **Chapter 1,** it is also difficult to define overall meat quality because it is a mix of sanitary, nutritional, technological and organoleptic components. Moreover, the meaning of quality of carcasses changes in several world areas according to the local customs (Rubino *et al.*, 1999). Therefore, it becomes necessary to investigate individual components of meat quality, such as visual aspects of meat quality, *e.g.* the colour of both fat and lean, or eating quality. Eating quality refers to the sensory attributes of meat, including toughness, juiciness and flavour, which in turn are affected by meat quality traits such as pH (Watanabe *et al.*, 1996) or fatty acid composition (Fisher *et al.*, 2000). Additionally, as many of the meat quality traits are

49

interrelated, it is important that both the properties of individual traits and the interrelationships between these traits are understood, so that gains in one trait will not be offset by a decline in another trait.

This **Chapter** investigates quantitative genetic aspects of carcass and meat quality traits in Scottish Blackface sheep and has three main aims. First, to estimate the heritability of carcass composition, as measured by CT, and meat quality traits; second, to calculate genetic relationships between these traits, and consequently investigate the implications of genetically altering carcass fatness; third, to investigate the prediction of meat quality traits using live animal measurements.

3.2 Materials and Methods

3.2.1 Animal population

The population studied throughout this thesis (**Chapters 3, 4, 5, 6, and 7**) comprised purebred Scottish Blackface lambs born during the period 2000 to 2003, at Roslin Institute's Blythbank farm. These lambs were derived from LEAN and FAT lines of Blackface sheep, previously divergently selected for predicted carcass lean proportion. A description of the selection procedures and initial responses to selection is given by Bishop (1993). Divergent selection ceased in 1996, after which the lines were maintained as closed populations with no further selection. The flock consisted of 200 ewes, split almost equally between the LEAN and FAT lines. A small proportion of reciprocal LEAN x FAT line crosses was made at the 1999 matings, so that a cohort of F_1 lambs were born in April 2000 along with a majority of purebred LEAN and FAT line lambs. The male F1 lambs were then backcrossed to the purebred LEAN and FAT line ewes to create a population of F_1 x LEAN and F_1 x FAT lambs from 2001 to 2003, for the purpose of quantitative trait loci (QTL) detection. A small number of F_1 x (F_1 x LEAN) and F_1 x (F_1 x FAT) were born in 2003 when female backcross lambs from 2001 themselves became dams. Additionally, F_1 females born in 2000 gave rise to a small number of F_2 progeny. Details are given below regarding the animals that were phenotyped.

Standard husbandry procedures were applied in this flock; all lambs were tagged at birth, with parentage, day of birth, sex and mortalities recorded. Each year the lambs were kept in two different groups, *i.e.* on two separate fields, for ease of management. Parentage information was maintained for all animals born after 1986, giving a total of 4847 known animals in the pedigree. In this dataset, a total of 23 sires produced phenotyped progeny.

3.2.2 Phenotypic measurements

3.2.2.1 *In vivo* carcass composition

Computer X-ray tomography (CT) was used to obtain non-destructive *in vivo* estimates of the carcass composition on 700 lambs, at 24 ($\pm$ 2) weeks of age, with scanning taking place, each year, in a 3-day period. Carcass composition was assessed on 100 lambs in year 2000 (50 per line) and *ca.* 200 lambs in years 2001 and 2003, with approximately equal number of males and females per year. Cross-sectional scans were taken at the ischium (ISC), the 5th lumbar vertebra (LV5) and the 8th thoracic vertebra (TV8), and from each scan image, the areas and image densities were obtained for the fat, muscle and bone components of the carcass.

From these measurements along with live weight (LW), a number of further traits were estimated, using prediction equations for total weights of each tissue that had been previously derived from a calibration study on unrelated Scottish Blackface lambs of the same age, as follows:

Fat weight (fatwt)= -1340 + (63.6 X LW) + (0.351 X Fat area LV5)

+ (0.248 X Fat area TV8) (R^2= 91.9%);

Bone weight (bonwt)= -102 + (39.1 X LW)+(0.241 X Bone area ISC)

+ (0.762 X Bone area LV5) (R^2= 73.5%);

Muscle weight (muswt)= -1640 + (87.3 X LW) + (0.242 X Muscle area ISC)

+ (0.163 X Muscle area TV8) (R^2= 84.8%);

Total carcass weight (cs (tot) wt)= Fat weight + Bone weight + Muscle weight;

Killing Out proportion (KO)= Total carcass weight/LW;

Fat proportion (Fprop) = fatwt/cs (tot) wt;

Muscle proportion (Mprop)= muswt/cs (tot) wt;

Bone proportion (Bprop) = bonwt/cs (tot) wt;

Muscle to Bone ratio (M: B)= muswt/bonwt;

3.2.2.2 Meat quality traits

Meat quality measurements, by definition, require the destruction of the animal, thus these measurements were only performed on male animals. Females were retained to maintain the flock. Measurements were performed on 25 LEAN and 25 FAT line male lambs in year 2000 and on 100, 100 and 99 male lambs in 2001 to 2003, at *ca.* 8 months of age, on lambs on which CT carcass composition measurements were previously obtained.

Measurements were performed at the University of Bristol, on cohorts of 20 animals treated identically during their growth, transportation and pre-slaughter periods. The lambs were slaughtered at the Bristol University abattoir, Langford, and were electrically stunned and conventionally dressed.

Carcass and meat quality measurements included pH at 45 min and 24 h post slaughter, colour of the meat, conformation score and fat classification, hot and cold carcass weight, toughness (or shear force), chemical composition and taste panel of cooked meat. Measurement protocols were as follows. In order to prevent cold-shortening of the muscles, the carcasses were hung at the ambient temperature for about 5 h prior to chilling at 1 °C. Twenty-four hours after slaughter, carcasses were classified for fat cover and conformation using 1 to 15 scales similar to those for beef carcasses described by De Boer *et al.* (1974). Muscle pH was measured using a glass electrode on 1g core of muscle, sampled from the *m. longissimus thoracis* at the last rib at 45 min (pH_{45}) and 24 h (pH_U) post slaughter and homogenised in 10 ml of ice cold iodoacetate buffer. Muscle colour was recorded using a Minolta Chromameter (Minolta CR-100, Minolta Camera Company, Milton Keynes, UK) employing the CIELAB colour space on loin chops cut at the time of boning, overwrapped with oxygen permeable film and allowed to bloom for 2 h before taking an average of three measurements on the cut surface. Colour was measured with reference to reflectance (L^*: 0 =black, 100=white) and two colour co-ordinates, a^* and b^*, with the extreme colours of a^* equal to red (positive) and green (negative) and b^* equal to yellow (positive) and blue (negative). The hue angle, $\tan^{-1} (b^*/ a^*)$, describes the colour in two dimensional space and saturation (chroma), $\sqrt{a^{*2} + b^{*2}}$, describes the intensity of the colour (MacDougall and Rhodes, 1972). Muscle toughness was measured after 10 days conditioning of loin joints (*m. thoracis et lumborum* (LTL)) at 1°C and then frozen and stored at −20ºC until analysis. Samples were thawed overnight at 4ºC, cooked in a water bath at 80°C until they reached an internal centre temperature of 78°C, cooled in running water and left in ice-water overnight. Blocks of muscle (7 to 10) were cut 10 x 10 x 20 mm with fibre direction parallel to the long axis. Muscle blocks were sheared at right angles to the fibre direction on a Stevens CR Analyser tensile testing machine using Volodkevich-type jaws, which recorded the shear force at first yield.

Proximate chemical analyses were carried out on the raw *m. semimembranosus* from the left leg. Frozen muscle samples were cut into smaller portions, minced three times through a 2-mm plate to ensure homogeneity, and chemically analysed. Total proportions of moisture and protein were determined according to the Association of Official Analytical Chemists methods (AOAC, 1997). Protein content was determined by the block digestion method. Moisture content was determined by drying at 100°C for 24 h. Lipids were extracted from 10g duplicate samples of lean muscle essentially as per Folch *et al.* (1957), separated into neutral and phospholipid, saponified, methylated, and individual fatty acids separated by column chromatography and quantified as described by Demirel *et al.* (2004). Total fatty acid was taken as the sum of all the phospholipid and neutral lipid fatty acids quantified, and this total was used as an estimate of intramuscular fat content.

Descriptive sensory analyses were performed using a trained taste panel. A section of LTL was removed 24 h after slaughter from the left side of the carcass, was packaged under vacuum and conditioned at 1°C for a further 10 days when it was frozen at -20°C prior to assessment of eating quality under standardized conditions. Samples were thawed at 4°C overnight and cut into 2.5 cm chops, which were grilled to an internal temperature of 78°C, as measured by a thermocouple inserted into the centre of the muscle. Eating quality traits analysed in this **Chapter,** included intensities of toughness, juiciness and lamb flavour (the remaining eating quality traits will be analysed in **Chapter 6**). Overall liking, a hedonic characteristic, was also assessed. Ten experienced taste panellists assessed the eating-quality of every animal on a 100mm unstructured line scales, where 0= nil, 100= extreme for each eating quality attribute. Specifically, for juiciness 0= not juicy, 100= extremely juicy, for toughness 0= very tender, 100= very tough, for flavour 0= no lamb flavour, 100= full lamb flavour and for overall liking 0= least favourable impression, 100= most favourable impression.

3.2.3 Statistical analysis

3.2.3.1   Data summary and trait definition

Data were initially analysed using regression in order to identify significant fixed effects, covariates and two-way interactions, using Genstat (2003). Prior to analysis the selection lines were coded as follows: 1= LEAN, 2= FAT, 3= F_1 x LEAN, 4= F_2, 5= F_1 x FAT, 6= F_1 x (F_1 x LEAN) and 7= F_1 x (F_1 x FAT). Multiple birth lambs comprised 0.52 of the data. The analysis of CT traits included fixed effects of line category (seven classes as given above), lamb sex (male or female), dam age (four classes: 2, 3, 4 and 5 years old), year of birth of the lambs born (four classes: 2000, 2001, 2002 and 2003), management group (1 or 2) and litter size (two classes: 1 or 2). Triplets comprised proportionately less than 0.02 of the data and were combined with the twin lambs in the classification of litter size. The model for meat quality traits included line category (seven classes), dam age (four classes), litter size (two classes) and year by slaughter day of lamb (15 classes). The only interaction found to be significant was between fixed effects year and group. Date of birth was fitted as a covariate for the CT traits, as proxy for age at scanning. Fixed effects and covariates found to be significant in the multiple linear regression analysis were included in the subsequent variance component analyses for each trait.

To analyse post-slaughter fat classification and conformation score values, two new traits were constructed (fat class and conformation value) as presented in **Chapter 2**. Additionally, subcutaneous fat content (g/kg) was estimated from fat classification using the following transformation, *i.e.* 1 → 40, 2 → 80, 3L → 110, 3H → 130, 4L → 150, 4H → 170 and 5 → 200 (Kempster *et al.*, 1986).

3.2.3.2   Estimation of fixed effect means and variance components

Residual maximum likelihood methods were used to estimate variance components using an animal model, fitting the complete pedigree structure (4847 animals), using ASReml (Gilmour *et al.*, 2004). The model included selection line, the year by group interaction, litter size and age of dam as fixed effects. Date of birth was fitted as a covariate and the fixed effect of lamb sex was also included in the model for CT measurements.

The (co)variance components were obtained by three different animal models in each univariate single trait analysis. The first model fitted the animal direct genetic effect as a random effect (model 1: $y= Xb + Za + e$). The second model (model 2: $y= Xb + Za + Sc + e$), where the effect of common environment was estimated, fitted litter as a further random effect, where y is the vector of observations on the specific trait of the animal; b is the vector of fixed effects, a is the vector of additive random animal (genetic) effects, c is the vector of random common environmental effects (litters) and e is the vector of random residual effect. X, Z and S are the incidence matrices relating records to fixed, animal and common environmental effects respectively. Model 2 was compared against model 1 using likelihood ratio test. Additionally, a model including maternal genetic effects was tested, however this never fit significantly better than model 2. Hence, results from these analyses are not tabulated.

It was assumed that the expectations (E) of the variables were: $E(y)=Xb$; $E(a)=0$; $E(e)=0$; $E(c)=0$ and that common environmental effects and residuals effects are independently distributed with means of zero and variances σ_c^2 and σ_e^2 respectively. Therefore, $var(a)=A\sigma_a^2$; $var(c)= I\sigma_c^2$; $var(e)=I\sigma_e^2$, where A is the numerator relationship matrix of animals in the model; I is the identity matrix; σ_a^2 is the additive genetic variance for direct effects (animal), and σ_c^2 and σ_e^2 are defined above.

Correlations between traits were estimated from bivariate analyses, fitting the animal, litter and residual effects. Covariances were fitted between animal effects and between residuals. Approximate standard errors for heritabilities and correlations were constructed by ASReml from approximation to variances of ratios and products. Correlations were estimated between CT measurements and, based on results presented below, between muscle density, fat density and traits assessed at slaughter.

3.2.3.3 Line effects
In order to compare the differences between lines for each measured trait, true line effects were estimated as the generalised least squares solutions to equations describing the genetic composition of the 7 line categories *i.e.* **LINE= $(X'V^{-1}X)^{-1}X'V^{-1}Y$**, where **LINE** is a vector of line solutions (LEAN and FAT) for each trait, **Y** is a vector of predicted means for the 7 line categories, **V** is the variance/covariance matrix of these solutions, and **X** is the incidence matrix relating genetic line to the 7 line categories. Standard errors of the line means and differences were then constructed from the appropriate elements of $(X'V^{-1}X)^{-1}$. Predicted means and variances for line categories were estimated using Genstat (2003) residual maximum likelihood (REML) procedure.

3.2.3.4 *In vivo* prediction of meat quality traits
Meat quality traits, as described in this study, require the destruction of the animal. Thus, it would be advantageous to be able to predict these traits using live-animal measures. It was hypothesised that some of the meat quality measures may be predictable from attributes of the CT scan image, particularly the tissue densities. Multiple regression analyses were performed on precorrected standardized data (mean= 0, standard deviation= 1) to investigate this hypothesis. In all cases the dependent variable was a meat quality trait, and the independent variables investigated were live weight and the areas and densities of each tissue at each site, as assessed by CT.

Multiple regression prediction models, with each possible subset evaluated by single stepwise addition and elimination were developed. Goodness of fit was assessed by the

Mallow's Cp criterion (Mallows 1973), with this statistic being used to avoid over-parameterising the model. The statistic is: $C_p = SSR_p/s^2_\varepsilon - (n-2p)$, where SSR_p is the residual sum of squares from a model with p parameters (including β_0), and s^2_ε is the mean square error from the regression equation with the largest number of independent variables. The best-fitting model should have $C_p \approx p$.

Prediction equations were derived on the data from the 2000, 2001 and 2002-born cohorts of lambs, and these equations were then used to predict the meat quality traits for the 2003-born lambs. The predicted scores were then correlated with the observed scores on the dependent variable ($r_{yy'}$); this is the cross-validity coefficient. The smaller the difference between the original *R*-squared and $r_{yy'}^2$, the greater the confidence we can have in the generalizability of the equation. The final step of the validation procedure was to combine both datasets and create a final prediction equation based on the whole dataset.

3.3 Results

3.3.1 Summary Statistics

Summary statistics for the CT assessed traits are shown in Table 3.1, with significant (*P*< 0.05) line differences shown in bold. The lines differed in the expected direction (*i.e.* FAT line fatter) for predicted fat proportion, fat areas, fat areas scaled by live weight and predicted weight of fat. Additionally, at two of the three measurement sites, the FAT line had significantly less dense muscle, indicative of a greater intramuscular fat content, and had significantly more dense bone. Counteracting the changes in fatness, the LEAN line animals had greater areas, weights and proportions of muscle and bone.

Table 3.1 *Predicted line means[‡], trait standard deviations, and line differences (with standard error) for CT traits. Significant (P< 0.05) line differences are shown in bold*

Trait	FAT Line	LEAN Line	S.D.	Line Difference (FAT-LEAN)	S.E. (Diff)
Live weight at CT (kg)	32.76	33.11	4.97	-0.35	0.52
Predicted bone proportion	0.19	0.20	0.02	-0.003	0.003
Predicted fat proportion	0.21	0.20	0.05	**0.014**	0.005
Predicted muscle proportion	0.59	0.60	0.03	**-0.011**	0.004
Average fat area[†] (mm^2/kg)	124	113	36.5	**10.9**	1.30
Fat area ISC[†] (mm^2/kg)	157	149	40.7	**8.40**	1.64
Fat area LV5[†] (mm^2/kg)	63.7	54.9	26.0	**8.81**	1.11
Fat area TV8[†] (mm^2/kg)	155	139	49.5	**16.0**	1.59
Fat area ISC (mm^2)	5021	4792	1450	**229**	8.50
Fat area LV5 (mm^2)	2140	1845	949	**295**	6.94
Fat area TV8 (mm^2)	5209	4628	1932	**581**	11.6
Fat density ISC	-70.1	-69.8	5.85	-0.25	0.54
Fat density LV5	-65.6	-64.5	6.15	-1.13	0.63
Fat density TV8	-68.9	-68.8	6.00	-0.10	0.66
Bone area ISC (mm^2)	2451	2504	3723	**-53.0**	5.28
Bone area LV5 (mm^2)	686	711	128	**-24.4**	2.40
Bone area TV8 (mm^2)	2763	2894	527	**-131**	5.76
Bone density ISC	326	325	39.3	1.80	1.81
Bone density LV5	369	360	52.2	**8.30**	1.68
Bone density TV8	321	315	39.2	**6.10**	1.39
Muscle area ISC (mm^2)	19971	20493	2061	**-522**	11.7
Muscle area LV5 (mm^2)	6772	6859	847	**-87.0**	7.54
Muscle area TV8 (mm^2)	9334	9398	1241	**-64.0**	9.86
Muscle density ISC	42.9	43.1	2.85	-0.24	0.40
Muscle density LV5	44.2	45.7	3.08	**-1.45**	0.45
Muscle density TV8	42.4	44.2	5.31	**-1.84**	0.58
Predicted fat weight (kg)	2.82	2.60	1.01	**0.216**	0.007
Predicted bone weight (kg)	2.51	2.56	3.21	**-0.052**	0.004
Predicted muscle weight (kg)	7.54	7.72	1.09	**-0.177**	0.008
Carcass total weight (kg)	12.9	12.9	2.21	-0.01	0.01
Muscle: Bone ratio	3.01	3.02	0.21	-0.01	0.13
Killing Out proportion	0.393	0.390	0.03	0.003	0.004

† Fat area scaled by live weight.
‡ Line means predicted from the 7 line categories.

Summary statistics for traits measured at slaughter and the major meat quality traits are shown in Table 3.2. The lines differed in the expected direction for subcutaneous fatness (*i.e.* FAT line fatter), but this was accompanied by an unexpected difference in live weight at slaughter, with the FAT line being less heavy. Non-significant trends in the same direction were also seen for hot and cold carcass weight. The other significant line differences were seen for colour attributes (colour L[*] and hue) and juiciness, with FAT line meat being significantly lighter and yellower and perceived to be more juicy than the LEAN line. Finally, the FAT line had a significantly greater intramuscular fat content than the LEAN line, as previously suggested by the muscle density line differences.

Table 3.2 *Predicted line means[†], trait standard deviations, and line difference (with standard error) for traits measured at slaughter and meat quality assessments. Significant (P< 0.05) line differences are shown in bold*

Trait	FAT Line	LEAN Line	S.D.	Line Difference (FAT-LEAN)	S.E. (diff)
Slaughter Live Weight (kg)	37.0	39.2	5.82	**-2.22**	1.15
Cold Carcass Weight (kg)	17.3	17.9	3.14	-0.64	0.65
Hot Carcass Weight (kg)	17.7	18.4	3.22	-0.69	0.68
Fat class value (units)	2.19	1.97	0.48	0.224	0.29
Conformation value (units)	3.77	3.51	0.62	0.26	0.22
Subcutaneous fat (g/kg)	93.0	80.7	22.7	**12.3**	2.34
Shear force (kg)	5.34	5.14	2.21	0.20	0.58
Redness (a*) (+ve→red)	17.0	17.3	2.43	-0.26	0.41
Yellowness (b*) (+ve→yellow)	8.03	7.75	1.32	0.28	0.37
Lightness (L*) (0=black, 100=white)	41.5	40.3	3.24	**1.11**	0.42
Hue (0°=red, 90°=yellow)	25.3	24.0	4.06	**1.30**	0.64
Saturation	18.9	19.0	2.45	-0.11	0.44
pH_{45}	6.67	6.73	0.17	-0.06	0.17
Ultimate pH	5.73	5.71	0.10	0.018	0.14
Juiciness (units)	43.4	40.7	7.10	**2.73**	1.14
Toughness (units)	36.3	36.4	10.3	-0.05	1.16
Overall Liking (units)	22.0	22.0	7.12	0.06	1.22
Flavour (units)	25.2	26.9	6.39	-1.66	1.04
Dry matter proportion	0.252	0.250	0.009	0.002	0.003
Nitrogen proportion	0.033	0.033	0.001	-0.000	0.001
Intramuscular fat (mg/100g muscle)	2507	2426	802	**81.0**	7.89

† Line means predicted from the 7 line categories.

3.3.2 Genetic parameters

Univariate heritability estimates and litter effects for carcass traits are shown in Table 3.3. Common environmental effects (litter) were significant, and therefore fitted, for most traits. It is recognised that the litter effect will most probably contain a maternal genetic component, however the structure of this dataset is unsuitable for estimating maternal genetic effects and, when tested, the model containing maternal genetic effects always gave a poorer fit than that with litter effects.

Almost without exception, live weight and the raw CT measurements, *i.e.* weights and densities of tissues at the three sites, were moderately to highly heritable. For example, heritabilities for fat areas ranged from 0.50 to 0.76, whilst litter effect estimates were lower but still substantial at 0.24 to 0.33. Only for muscle and bone areas were heritabilities consistently below 0.4. Litter effects were significant for many of these measurements, the exceptions being the three fat and muscle densities, fat area TV8 and bone density TV8. Heritabilities for the derived tissue weight traits were less consistent. However, in the two cases where the heritabilities were low, bone weight and KO, unexpectedly large litter effects were seen.

Table 3.3 *Univariate heritabilities (h^2), litter effects (c^2) both with standard errors (s.e.) and variance components[†] for carcass traits assessed by CT*

Trait	h^2	S.E. (h^2)	c^2	S.E. (c^2)	σ_p^2
Live weight at CT (kg)	0.41	0.15	0.32	0.09	19.45
Fat area ISC (mm^2)	0.50	0.13	0.33	0.09	1.57
Fat area LV5 (mm^2)	0.66	0.22	0.24	0.18	0.75
Fat area TV8 (mm^2)	0.76	0.08	-	-	3.29
Fat density ISC	0.48	0.10	-	-	24.93
Fat density LV5	0.46	0.11	-	-	37.24
Fat density TV8	0.63	0.09	-	-	31.74
Bone area ISC (mm^2)	0.23	0.09	0.06	0.07	73760
Bone area LV5 (mm^2)	0.35	0.11	0.10	0.06	8320
Bone area TV8 (mm^2)	0.49	0.10	0.19	0.06	0.17
Bone density ISC	0.39	0.12	0.21	0.06	788.4
Bone density LV5	0.49	0.12	0.28	0.07	1289
Bone density TV8	0.51	0.09	-	-	620.4
Muscle area ISC (mm^2)	0.33	0.11	0.12	0.06	4.41
Muscle area LV5 (mm^2)	0.32	0.12	0.24	0.06	0.73
Muscle area TV8 (mm^2)	0.34	0.12	0.21	0.06	1.40
Muscle density ISC	0.82	0.08	-	-	7.51
Muscle density LV5	0.45	0.10	-	-	7.502
Muscle density TV8	0.34	0.10	-	-	9.24
Predicted fat weight (kg)	0.60	0.28	-	-	0.96
Predicted bone weight (kg)	0.14	0.11	0.57	0.08	3.62
Predicted muscle weight (kg)	0.48	0.17	0.27	0.09	0.96
Carcass total weight (kg)	0.55	0.20	-	-	4.20
Muscle: Bone ratio	0.56	0.18	0.21	0.09	0.38
Killing Out proportion	0.01	0.06	0.88	0.04	0.21

[†] Results presented for the best fit model, as assessed by the likelihood ratio test.

The estimated genetic and phenotypic correlations for the carcass traits measured on live animal by CT, estimated using bivariate analysis are shown in Table 3.4. The heritability values presented in Table 3.3 (using univariate models) and Table 3.4 (using the average of the bivariate heritability estimates) are very similar. Firstly consider genetic correlations between the same measurements taken at different sites on the carcass. Genetic correlations among muscle densities were positive, but the ISC measurement was only weakly correlated with the other two. In contrast, the genetic correlations among the three muscle areas traits were strong and positive, with an average of 0.93. Additionally, the genetic correlations between the three fat densities and between the three fat areas were positive and very high, averaging 0.84 and 0.96 respectively. Hence, muscle areas, fat areas and fat densities in different parts of the carcass may be considered genetically the same trait.

Now consider correlations between different tissue measurements. The estimated genetic correlations between muscle area and muscle density traits were negative and generally low to moderate. In contrast, fat area and fat density were moderate negatively genetically correlated. Genetic correlations between fat area and muscle area were moderately positive. In addition, fat density and muscle density were moderate positively correlated. On the other hand, muscle density and fat area were moderate negatively genetically correlated, which could be interpreted as indicating that muscle density is an indicator of intramuscular fat content, hence carcass fatness in general. Muscle area and fat density were moderate negatively correlated.

Table 3.4 Heritabilities, genetic and phenotypic correlations (with standard errors) for CT traits†

	Muscle Density ISC	Muscle Density LV5	Muscle Density TV8	Muscle Area ISC	Muscle Area LV5	Muscle Area TV8	Fat Density ISC	Fat Density LV5	Fat Density TV8	Fat Area ISC	Fat Area LV5	Fat Area TV8
Muscle Density ISC	**0.81(0.08)**	0.26(0.04)	0.23(0.04)	-0.15(0.04)	-0.13(0.05)	-0.19(0.04)	0.42(0.04)	0.35(0.04)	0.31(0.04)	-0.52(0.03)	-0.40(0.04)	-0.41(0.04)
Muscle Density LV5	0.13(0.14)	**0.44(0.10)**	0.64(0.02)	-0.18(0.04)	-0.17(0.04)	-0.24(0.04)	0.35(0.04)	-0.13(0.04)	0.33(0.04)	-0.48(0.03)	-0.51(0.03)	-0.52(0.03)
Muscle Density TV8	0.29(0.15)	0.93(0.09)	**0.35(0.10)**	-0.21(0.04)	-0.25(0.04)	-0.17(0.04)	0.27(0.04)	-0.15(0.04)	0.29(0.04)	-0.42(0.04)	-0.42(0.04)	-0.49(0.03)
Muscle Area ISC	-0.41(0.17)	-0.19(0.20)	-0.45(0.21)	**0.28(0.10)**	0.74(0.01)	0.69(0.02)	-0.30(0.04)	-0.12(0.04)	-0.29(0.04)	0.52(0.03)	0.48(0.03)	0.51(0.03)
Muscle Area LV5	-0.24(0.15)	0.01(0.19)	-0.47(0.19)	0.96(0.09)	**0.32(0.12)**	0.77 (0.01)	-0.27(0.04)	-0.06(0.04)	-0.23(0.04)	0.47(0.04)	0.40(0.04)	0.43(0.04)
Muscle Area TV8	-0.31(0.15)	-0.22(0.19)	-0.16(0.22)	0.87(0.07)	0.96(0.09)	**0.35(0.10)**	-0.29(0.04)	-0.07(0.04)	-0.25(0.04)	0.50(0.03)	0.44(0.04)	0.51(0.03)
Fat Density ISC	0.65(0.11)	0.55(0.15)	0.43(0.18)	-0.94(0.09)	-0.65(0.16)	-0.65(0.14)	**0.45(0.09)**	0.55(0.03)	0.64(0.02)	-0.68(0.03)	-0.68(0.02)	-0.68(0.02)
Fat Density LV5	0.78(0.11)	0.06(0.21)	-0.03(0.22)	-0.64(0.22)	-0.60(0.20)	-0.54(0.20)	0.80(0.08)	**0.45(0.10)**	0.55(0.03)	-0.37(0.04)	-0.51(0.03)	-0.42(0.04)
Fat Density TV8	0.47(0.11)	0.69(0.13)	0.53(0.15)	-0.65(0.18)	-0.47(0.17)	-0.49(0.16)	0.92(0.06)	0.80(0.08)	**0.60(0.09)**	-0.64(0.03)	-0.64(0.03)	-0.75(0.02)
Fat Area ISC	-0.59(0.09)	-0.46(0.13)	-0.50(0.14)	0.93(0.09)	0.40(0.04)	0.73(0.10)	-0.89(0.06)	-0.82(0.11)	-0.82(0.07)	**0.50(0.13)**	0.82(0.01)	0.85(0.01)
Fat Area LV5	-0.49(0.09)	-0.62(0.11)	-0.55(0.13)	0.73(0.14)	0.41(0.14)	0.57(0.13)	-0.88(0.06)	-0.76(0.10)	-0.82(0.07)	0.98(0.09)	**0.65(0.12)**	0.90(0.01)
Fat Area TV8	-0.48(0.09)	-0.51(0.11)	-0.62(0.11)	0.75(0.14)	0.35(0.13)	0.58(0.12)	-0.89(0.05)	-0.85(0.09)	-0.93(0.04)	0.93(0.03)	0.97(0.09)	**0.75(0.08)**

† Heritabilities on diagonal, phenotypic correlations above and genetic correlations below diagonal.

Heritabilities for traits assessed at slaughter and meat quality traits are shown in Table 3.5. Maternal and litter effects were generally not significant. These heritabilities were less well estimated than those for the CT traits, simply because of the smaller available dataset. In general terms, a dataset of *ca.* 350 phenotypic observations with nine sires contributing most progeny would be inadequate for genetic parameter estimation. However, genetic parameters are estimable from this dataset with an acceptable degree of precision in part because of the large and complex pedigree available. Most traits were moderately heritable and, in the case of the carcass classification measures (fat class and conformation), surprisingly heritable. These results indicate that genetic change, by some means, should be feasible for most traits investigated. A factor which may contribute to the larger than expected heritabilities for fat classification and conformation score is the experimental design, in which animals were slaughtered at a fixed age rather than a fixed degree of 'finish', as is current commercial practice in the UK sheep industry.

Table 3.5 *Heritabilities for traits assessed at slaughter and meat quality traits*

Trait	h^2	S.E.(h^2)	σ_p^2
Slaughter live weight (kg)	0.30	0.20	27.1
Cold carcass weight (kg)	0.47	0.19	7.41
Hot carcass weight (kg)	0.47	0.19	7.12
Fat class value (units)	0.33	0.16	0.21
Conformation value (units)	0.52	0.18	0.28
Subcutaneous fat (g/kg)	0.34	0.16	481
Shear force (kg)	0.39	0.16	3.84
Redness (a*) (+ve→red)	0.45	0.19	2.17
Yellowness (b*) (+ve→yellow)	0.33	0.17	1.02
Lightness (L*) (0=black, 100=white)	0.15	0.12	3.43
Hue (0°=red, 90°=yellow)	0.30	0.15	3.96
Saturation	0.45	0.18	2.78
pH_{45}	0.54	0.18	0.03
Ultimate pH	0.21	0.14	0.01
Juiciness (units)	0.21	0.12	35.1
Toughness (units)	0.15	0.13	99.7
Overall liking (units)	0.22	0.13	51.7
Flavour (units)	0.11	0.11	23.8
Dry matter proportion	0.51	0.16	0.98
Nitrogen proportion	0.01	0.06	0.01
Intramuscular fat (mg/100g muscle)	0.32	0.09	664000

Results from the analyses investigating *in vivo* prediction of meat quality traits (see below) indicated that muscle density measurements were informative in terms of predicting some meat quality traits. Therefore, phenotypic and genetic correlations were estimated between average muscle density and carcass and meat quality traits, and the results are presented in Table 3.6. Genetic correlations involving redness, yellowness, lightness of meat and saturation were all less than 0.10 and are not presented. Phenotypically, most of the correlations with average muscle density shown in Table 3.6 were low to moderate, although those with live and carcass weights, fatness traits, juiciness, flavour and overall liking were all negative and significantly different from zero. Genetic correlations tended to be stronger than the phenotypic correlations, although interpretation of some correlations is hindered by large standard errors. Muscle density is strongly negatively genetically correlated with fatness traits, including intramuscular fat content, and it is also strongly negatively correlated with juiciness, flavour, overall liking and muscle dry matter content.

Table 3.6 *Phenotypic and genetic correlations (with standard errors) for meat and carcass traits with average muscle density*

Trait	Phenotypic correlation	Genetic correlation
Cold Carcass	-0.23 (0.07)	-0.30 (0.21)
Hot carcass	-0.25 (0.07)	-0.36 (0.20)
Live Weight	-0.20 (0.08)	-0.34 (0.30)
Fat class value	-0.26 (0.07)	-0.67 (0.19)
Conformation value	0.10 (0.07)	0.38 (0.19)
Subcutaneous fat	-0.24 (0.07)	-0.62 (0.19)
Shear force	-0.16 (0.06)	-0.49 (0.14)
Hue angle	-0.05 (0.07)	-0.31 (0.27)
pH_{45}	0.06 (0.07)	0.30 (0.39)
pH Ultimate	0.14 (0.07)	0.41 (0.31)
Dry matter proportion	-0.56 (0.04)	-0.87 (0.09)
Intramuscular Fat	-0.57 (0.04)	-0.67 (0.14)
Toughness	0.15 (0.07)	-0.08 (0.39)
Juiciness	-0.16 (0.07)	-0.71 (0.22)
Overall liking	-0.29 (0.05)	-0.80 (0.21)
Flavour	-0.20 (0.07)	-0.73 (0.24)

In addition, genetic and phenotypic correlations were estimated between average fat density and carcass and meat quality traits. Average fat density was strongly correlated with weight traits (genetic correlations with cold carcass weight and live weight were 0.92 and 0.83, respectively), but correlations for all the other traits were not significantly different from zero. Attempts to estimate genetic correlations with other CT traits resulted in either failed to converge or had very large s.e.s, reflecting the weakness of the meat quality dataset for estimating genetic correlations.

Finally, due to the observation that intramuscular fat content was significantly correlated with average muscle density and is an important predictor of meat quality, the genetic and phenotypic correlations were estimated between intramuscular fat content and meat traits. Table 3.7 shows that intramuscular fat is moderately correlated with shear force, juiciness and flavour.

Table 3.7 *Phenotypic and genetic correlations (with standard errors) for intramuscular fat content with meat traits*

Trait	Phenotypic correlation	Genetic correlation
Shear force	-0.09 (0.07)	-0.54 (0.24)
Juiciness	0.12 (0.06)	0.69 (0.24)
Flavour	0.20 (0.06)	0.52 (0.22)

3.3.3 *In vivo* prediction of meat quality traits

Regression analysis was used to investigate whether or not meat quality traits could be predicted using CT measurements (*i.e.* densities and areas of fat, muscle and bone), using data collected from 2000 to 2002. Most traits were not well predicted, with the prediction equations explaining proportionately less than 0.1 of the trait variation. However, prediction was successful for some traits, and Table 3.8 shows results obtained for redness, juiciness, intramuscular fat, fat classification score and ultimate pH of lamb meat. In all cases where CT measures gave an adequate prediction of meat quality traits, it was muscle density that was the predominant predictor. Redness (a*) was accurately predicted (proportionately 0.5 of variance explained) and intramuscular fat, fat class value, ultimate pH and juiciness were also predicted with reasonable accuracy. Cross validation was performed using data collected on 2003-born lambs, and the correlations of predicted with observed scores were somewhat lower than those obtained from the regression analyses. The difference between the original R^2 and r_{yy}^2 was 0.01, 0.05 and 0.03 for the fat classification score, intramuscular fat and ultimate pH respectively, which indicates that the prediction equations were robust. However, for redness and juiciness of meat the difference was 0.20 and 0.31 respectively.

Table 3.8 Coefficients of determination (R^2) and cross-validity coefficient ($r_{yy'}$) for best-fit regression equations developed to predict in vivo meat quality traits from CT traits

Trait	Factors in best-fit model	Original R^2	$r_{yy'}$ (2000-2002 data)	$r_{yy'}^2$ (2003 data)	R^2- $r_{yy'}^2$
Redness (a*)	Muscle density (ISC, LV5, TV8)	0.50	0.71	0.30	0.20
Intramuscular fat	Muscle density (ISC, LV5, TV8)	0.25	0.50	0.20	0.05
Juiciness	Muscle density (TV8), Fat area (LV5)	0.34	0.58	0.03	0.31
Fat Class	Muscle density (LV5, ISC, TV8)	0.25	0.50	0.24	0.01
Ultimate pH	Muscle density (TV8, ISC), Fat area (LV5)	0.17	0.41	0.14	0.03

Prediction equations developed using the whole dataset are shown in Table 3.9, along with the R^2 values and residual standard deviations for each equation, as indicators of the precision of each equation. Redness (a*), intramuscular fat, fat class and ultimate pH were somewhat better predicted on the whole dataset than on the 2000 to 2002 subset.

Table 3.9 Final prediction equations of meat quality traits from CT traits with coefficient of determination (R^2) and standard error of estimate (RSD), based on whole dataset[†]

Final Prediction equations	R^2	RSD
Redness (a*)= 0.08 –0.19 x (Muscle density ISC) +0.22 x (Muscle density LV5) -0.71 x (Muscle density TV8)	0.62	0.79
Intramuscular fat= 9940 –120.4 x (Muscle density LV5) –29.09 x (Muscle density TV8) –14.2 x (Muscle density ISC)	0.33	6.62
Juiciness= -0.01+0.34 x (Muscle density TV8)+0.12 x (Fat area LV5)	0.31	0.95
Fat Class= 0.17-0.32 x (Muscle density LV5)-0.07 x (Muscle density ISC) -0.19 x (Muscle density TV8)	0.46	0.82
Ultimate pH= -0.11+0.24 x (Muscle density TV8)-0.08 x (Muscle density ISC) -0.24 x (Fat area LV5)	0.30	0.95

[†] For abbreviations see Table 3.1 footnote.

3.4 Discussion

The present study has produced novel and practically useful information on the genetic control of carcass and meat quality traits in Blackface sheep. As a broad summary, the results showed that reducing carcass fatness has simultaneously changed muscle density, intramuscular fatness, and changed aspects of muscle colour and eating quality traits (juiciness), making the muscle darker, less yellow and less juicy. The heritabilities observed for the meat quality traits indicate ample opportunities for altering most meat quality traits, provided that these traits can be adequately measured or predicted. The possibility of measuring these traits under field conditions is indicated by the predictions for redness, intramuscular fat, juiciness, fat class and ultimate pH made using CT measures, particularly aspects of muscle density. Together, these results, along with strong genetic correlations between muscle density and some of the meat quality traits, provide potential opportunities for genetically improving components of meat and carcass quality.

3.4.1 Line differences

Modest but significant FAT-LEAN line differences were seen for nearly all traits describing carcass fatness, with these traits being significantly higher in the FAT line. Furthermore, muscle densities (LV5, TV8) showed significant differences between lines, with the FAT line having lower muscle density than the LEAN line. This suggests more intramuscular fat in the FAT line, as fat is less dense tissue than muscle. Selection for carcass composition also changed some meat quality traits. The result that selection for increased leanness is likely to lead to a darker meat colour, which is generally less acceptable to consumers, has been seen in pigs (Cameron, 1990), although this difference is probably small compared to differences that may be induced by stress or nutritional factors. In the Cameron study, lines of Large White pig were selected for components of efficient lean growth rate, and a leaner carcass was achieved by reducing the rate of fat deposition. Lastly, the FAT line was significantly lighter in weight at slaughter (*ca.* 8 months) but not at the time of CT scanning (*ca.* 5 months), indicating that it is perhaps earlier maturing with a lower mature live weight than the LEAN line.

Bone densities (LV5, TV8) were found to be significantly higher in the FAT than the LEAN line. This result was in contrast with the study by Campbell *et al.* (2003), aimed at detecting the QTL for bone mineral density (BMD) in Coopworth sheep, in which bone density was generally negatively correlated with the amount of fat (subcutaneous, intramuscular, internal and total), the amount of muscle and body weight. Body weight and fat and muscle components are also correlated with bone density in humans (reported by Campbell *et al.*, 2003). Low BMD has been shown to be an important factor in osteoporotic fracture risk in humans (Cummings *et al.*, 1990). A number of QTL contributing to genetic variation in BMD were identified in the study by Campbell *et al.* (2003). These results along with the present study indicate that bone density is either correlated with, or a consequence of, body composition.

3.4.2 Inheritance of traits

The heritability estimates for CT traits on the live animal were moderate to high, indicating that CT provides a quick and reliable means of genetically changing carcass composition in sheep. In particular, heritabilities for tissue areas and densities at specific sites were nearly always high. There are not available other published estimates of site-specific genetic parameters, although Jones *et al.* (2004) report heritabilities for fat and lean weights derived from CT measurements. Their estimated heritabilities for lean and fat weight were 0.38 and 0.47, respectively, in general agreement with our heritability estimates of 0.60 and 0.48, respectively.

The closest comparisons for the CT measurement results are with ultrasonic measurements, which are also site-specific measurements. Genetic parameters for ultrasonic measurements have previously been reported in this and genetically related populations by Bishop (1993), using measurements taken in intensively housed lambs, and by Conington *et al.* (1995 and 2001), using measurements taken on lambs reared under hill conditions. Measurements taken under hill conditions tended to be less heritable than those taken under intensive conditions (0.27 and 0.30 *vs.* 0.36 for muscle depth, 0.16 and 0.25 *vs.* 0.39 for fat depth), and heritabilities were generally lower than those observed here from CT measures, particularly for fat measurements.

The heritability estimates for most of the meat quality traits measured in this study indicate that there is substantial genetic variation in these traits, such that meat quality can potentially be improved through direct selection on the traits or through indirect selection. Colour parameters were moderately heritable, with the additive genetic coefficients of variation being 0.07 for redness, 0.10 for yellowness and 0.03 for lightness indicating that genetic improvement for the first two colour attributes is potentially feasible. Although these results indicate that genetic improvement of lightness may be difficult, the lines did differ significantly. Breed differences have also been reported for sheep meat colour. For example, Küchtik *et al.* (1996) reported significant sheep breed differences in reflectance values of joints of muscle. Young *et al.* (1993) indicated that cooked Merino meat was darker than cooked Coopworth meat, and suggested that the darker colour was a result of higher iron content, present as myoglobin and other iron-containing proteins.

Due to a lack of information on heritabilities in meat traits in sheep, we will compare our results with published studies in other species. In pigs, the heritability estimates for muscle lightness, obtained by Lo *et al.* (1992) for Duroc and Landrace pigs and Larzul *et al.* (1997) for Large White pigs, were 0.11 and 0.20, respectively, similar to the estimate obtained in the present study. Hermesch *et al.* (2000) and Sellier (1998) found that muscle lightness in pigs was moderately inherited, with heritabilities of 0.29 and 0.28 respectively, higher than that of the presented study. The heritability estimate of muscle lightness (0.15) is almost identical to that obtained for pigs by Cameron (1990). There appear to be no published heritabilities for redness and yellowness.

The heritability for muscle pH at 45 min *post-mortem* was high (0.54) while the estimate for ultimate pH was much lower (0.21). However, the genetic and phenotypic correlations between pH_{45} and pH_u were 0.97 and 0.40 respectively, indicating that the rate and extent of the *post-mortem* pH fall are probably governed by the same genes. Our heritability for pH_u is similar to values seen in pigs, although our estimate for pH_{45} is higher than published values for pigs. For example, in the study of Hermesch *et al.* (2000), the heritability for pH_{45} in pigs was 0.15, *i.e.* somewhat lower than our value, and 0.14 for pH_u. Sellier (1988) found average heritability estimates of 0.16 and 0.21 for pH_{45} and pH_u, respectively, of pigs and Cameron (1990) reported an estimate of 0.21 for pH_u. However, heritability values alone can be misleading, as the variability in pH values is not large. In all cases cited above, the genetic coefficients of variation were less than 0.01, indicating that genetic change would be slow. Non-genetic factors affecting pH are well documented. For example, McGeehin *et al.* (2001) found that the most important factors that influence the pH of lamb meat were sex, carcass weight and to a lesser degree, age and ambient temperature. Also, rates of pH fall are faster in stressed animals, and ultimate pH is usually governed by diet. Care was taken in this experiment to minimise variations in pre-slaughter diet or stress levels.

Intramuscular fat content of the *semimembranosus* muscle was moderately heritable in this study ($h^2 = 0.32$), although this is slightly lower than values reported in pigs. Malmfors and Nilsson (1979) obtained heritability estimates of 0.58 and 0.68 for Swedish Landrace and Large White pigs, respectively. Scheper (1979) reported a heritability estimate of 0.35 for German Landrace pigs, and heritabilities of 0.55 were reported for Landrace pigs in Denmark (Just *et al.*, 1986) and in Switzerland (Schworer *et al.*, 1987). The average heritability weighted by number of sires for intramuscular fat content from previous reports is 0.53 (Sellier, 1988).

Previous published estimated heritabilities for taste panel evaluations of meat quality also come from pigs. Cameron (1990) reported heritability estimates for toughness, pork flavour, juiciness and overall liking of 0.23, 0.16, 0.18 and 0.16, respectively, from data on 40 full-sib litter groups of Duroc and halothane-negative British Landrace pigs, in agreement to estimates of 0.15, 0.11, 0.21 and 0.22 for these traits in the present study. In contrast, Lo *et al.* (1992) reported heritability estimates of 0.45, 0.13, 0.12 and 0.34 for toughness, pork flavour, juiciness and overall liking, respectively, from data on Duroc and Landrace pigs. Based on results of a small number of studies, eating quality traits in general seem to be low to moderately heritable of the order of 0.10 to 0.20 (Lo, 1990), in agreement with the present study. However, these estimates tend to be quite variable, as would be expected given the relatively small sample sizes generally used in these studies. Overall, these results indicate that sensory meat quality traits, assessed by taste panels, are determined to some extent by additive genetic effects and as such there is some scope for genetic improvement by means of selection.

3.4.3 Relationships between traits

Our dataset was large enough to allow estimation of genetic correlations between CT traits, and between some of the CT traits and some of the meat quality traits, but there was insufficient data for reliable estimation of correlations amongst the meat quality traits. As a broad summary of the CT traits, measurements of the attributes of the same tissue taken at different sites were generally strongly genetically correlated, indicating that they are essentially expressions of the same trait. Fat area was moderately positively genetically correlated with muscle area, but moderately to strongly negatively correlated with muscle and fat densities. Likewise muscle area was negatively correlated with muscle and fat densities. Genetic change in any of these attributes is feasible in principle, and correlated responses in other traits predictable from genetic theory.

There is no published information available on genetic relationships between carcass traits assessed by CT and meat quality traits in any species. We have demonstrated consistent relationships between muscle density and meat quality traits, however analyses involving other CT measurements generally did not give consistent or significant results. Muscle density was negatively genetically correlated with intramuscular fat, lamb juiciness, flavour and overall liking. Selection to decrease muscle density and hence increase intramuscular fat is expected to result in lamb meat that is juicier and with more flavour. Furthermore, muscle density was negatively correlated with shear force value (a quantitative measure of toughness), again highlighting a means of reducing toughness. The only other significant and consistent genetic correlation was between fat density and weight traits, with fat becoming denser as live weight increased.

Relationships of muscle density with meat quality traits were borne out when intramuscular fat content was investigated in more detail; it was positively correlated with juiciness and flavour, indicating that fat plays a role in these properties. Intramuscular fat content has long been recognised as a factor in eating quality, although the strength of the relationship has been questioned (Wood, 1990). De Vol *et al.* (1988) reported genetic correlation coefficients of 0.21, 0.32 and 0.23 for pork intramuscular fat content with juiciness, toughness and flavour, respectively. Based on a reassessment of these data, Lo *et al.* (1992) estimated correlations of 0.24 and 0.63 for intramuscular fat content and juiciness and flavour, respectively, and they suggested the relationship between eating quality and intramuscular fat content is dependent on the amount of intramuscular fat. In the same study it was found that selection for increased intramuscular fat content would decrease shear force value, as it was found in our study. According to Fernandes *et al.* (2002) intramuscular fat, extracted from *longissimus* muscle in beef, is a good indicator for flavour and juiciness: these authors observed a high heritability for intramuscular fat content (0.57) and strong genetic correlations between marbling and intramuscular fat (0.94 and 0.65). These results suggest there is a potential to improve meat quality for flavour and juiciness through intramuscular fat, and muscle density measures may be a means of doing this *in vivo*.

In vivo prediction of meat quality traits, at the phenotypic level, using CT measures was successful for some traits, including redness, intramuscular fat, juiciness, fat class and ultimate pH. In all cases, shown in Table 3.9, where CT measures predicted meat quality traits with an $R^2 > 0.3$, it was muscle density that was the predominant predictor, backing the results from the genetic correlations. However, it should be noted that using muscle density and hence intramuscular fat to improve juiciness, flavour and overall liking will result in an increase in intramuscular fatness, and this is in the opposite direction to consumers' perceived desire for decreased fatness.

3.5 Conclusions

This study has confirmed that CT measurements, in principle, provide an effective means of selecting for improved carcass composition, in agreement with the results of **Chapter 2**. Additionally, it has shown that there is considerable scope for genetic change in many meat quality traits, provided that they can be measured or predicted. This study has also shown that altering carcass fatness has simultaneously changed muscle density and intramuscular fatness, and changed aspects of muscle colour and eating attributes, with selection for increased fatness making meat lighter, more yellow and more juicy. Also, bone density (LV5, TV8) differed between lines, being higher in the FAT line and, together with other studies, shows that bone density is correlated with other traits, such as fat and muscle weight, indicating that it is an important trait for further research.

An important contribution made by this study is the realisation that it is possible to measure some meat quality traits under field conditions, using CT measures, particularly aspects of muscle density. Muscle density is a trait collected automatically during CT assessments of commercial animals, but currently not utilised. Therefore, this information could feasibly be utilised for predicting these traits and effecting changes in meat juiciness and flavour.

3.6 Implications

As a consequence of these experimental results and comparing them to previously published literature, it is obvious that CT traits provide more accurate measurement of carcass traits in the live animal than ultrasonics. Beyond doubt it is a very useful research tool. However, the expense and lack of portability of CT equipment, means that there are logistical difficulties in using it in breeding programs. But it is being used in the UK for two-stage selection strategies in which only elite animals, identified by other means, are CT scanned. Selecting effectively for improved meat quality attributes would require a similar approach, so that both the quantity of product and quality of product can be increased. Our results indicate the potential for a decrease in sheep eating quality resulting from selection for improved lean meat characteristics, and this must be avoided. Using field and CT measurements, multi-trait animal model evaluations could be introduced including traits such as muscle density that influence meat quality, and the resulting breeding values could be used in selection strategies that attempt to simultaneously improve lean meat output and improve meat quality characteristics.

Chapter Four
A partial genome scan to map quantitative trait loci for carcass composition, assessed by X-ray computer tomography, and meat quality traits

4.1 Introduction

In the past, leanness was considered one of the most important traits. As a result, dramatic improvements in body composition, mostly in pigs, have been made. However, it has been shown that lean meat is not always associated with good meat quality (Cameron 1990; Hovenier *et al.* 1992), and therefore several other traits must be considered to improve meat quality. Improving meat quality genetically is difficult by standard selection methods, but possible if, firstly, *in vivo* predictors could be used and secondly, if the genes responsible for meat quality are identified and mapped. A possibility of utilising CT measures, particularly muscle density, as predictors of certain meat quality attributes has been suggested in **Chapter 3**. However, even with this information, routine measurement of meat quality attributes will remain expensive and difficult.

Marker or gene-assisted selection has been suggested as a promising strategy for genetic improvement of difficult-to-measure traits such as meat quality (Meuwissen and Goddard, 1996). Identification of quantitative trait loci (QTL) for meat quality will be an important step towards marker-assisted selection. The development of genetic markers and their application to farm animals has progressed rapidly, opening new prospects for identifying chromosomal regions defining QTL. There is less activity in QTL identification in sheep populations compared to other livestock species, and surprisingly few QTL have been published for traits of direct relevance to meat production, apart from studies of individual major genes such as the callipyge locus (Freking *et al.*, 2002). Thus, there is a clear need to utilise genetic resources for research to detect QTL for carcass and meat quality traits in sheep. This **Chapter** addresses this issue and aims to identify QTL for carcass composition and meat quality traits.

4.2 Material and Methods

4.2.1 Animals and trait measurement

The population structure of this study has been provided in detail in **Chapter 3**. In summary, lambs were derived from LEAN and FAT lines of Blackface sheep, divergently selected for predicted carcass composition (Bishop, 1993). A small number of LEAN x FAT line crosses were produced in the 1999 matings, so that a cohort of F_1 lambs were born in April 2000 along with purebred LEAN and FAT line lambs. The male F_1 lambs were then backcrossed to the purebred LEAN and FAT line ewes to create a population of F_1 x LEAN and F_1 x FAT lambs from 2001 to 2003, for the purpose of QTL detection. This double backcross design created 9 half-sib families for QTL detection. On average, families contained 67 offspring for carcass composition traits measured on the live animal and 33 offspring for meat quality traits, with a range from 25 to 96 progeny and 12 to 46 progeny, respectively. A summary of the families is presented in Table 4.1.

Table 4.1 *Summary of the families used in the study for carcass traits assessed by CT and meat quality traits*

Family	Number of progeny for CT traits	Number of progeny for meat traits
S1	36	20
S2	88	38
S3	56	24
S4	53	30
S5	96	45
S6	25	12
S7	70	39
S8	82	46
S9	94	46

Phenotypic measurements of carcass composition, taken on cross-sectional scans at the ischium (ISC), the 5th lumbar vertebrae (LV5) and the 8th thoracic vertebrae (TV8), were obtained by computerised tomography (CT) on 600 5-month old live lambs. From each scan image the areas and image densities were obtained for the fat, muscle and bone components of the carcass. Additionally, meat quality measurements were made on 300 8-month old male lambs that had previously been CT scanned. Measurements were performed at the University of Bristol, on cohorts of 20 animals treated similarly during their growth, transportation and pre-slaughter periods. Meat quality measurements, presented in this study, included initial and final pH of *m. semimembranosus*, colour (L*, a* and b*) and sensory characteristics, as assessed by a trained taste panel (juiciness, lamb flavour, toughness and overall liking). Muscle pH was measured using a glass electrode on 1g core of muscle, sampled from the *m. longissimus thoracis* at the last rib and homogenised in 10 ml of ice cold iodoacetate buffer. Measurements were then taken at 45 min (pH_{45}) and 24 h (pH_U) post slaughter. Muscle colour was recorded on loin chops cut at the time of boning, overwrapped with oxygen permeable film and allowed to bloom for 2 h before taking an average of three measurements on the cut surface. Colour was measured with reference to reflectance (*L**: 0= black, 100= white) and two colour co-ordinates, *a** and *b**, with the extreme colours of *a** equal to red (positive) and green (negative) and *b** equal to yellow (positive) and blue (negative). Descriptive sensory analyses were performed using a trained taste panel. A section of *m. thoracis et lumborum* (LTL), removed 24h after slaughter from the left side of the carcass, was packaged under vacuum and conditioned at 1°C for a further 10 days, whereupon it was frozen at -20°C prior to assessment of eating quality under standardised conditions. Samples were thawed at 4°C overnight and cut into 2.5 cm chops, which were grilled to an internal temperature of 78°C, as measured by a thermocouple inserted into the centre of the muscle. Eating quality traits included intensities of toughness, juiciness and lamb flavour. Overall liking, a hedonic characteristic, was also assessed. Ten experienced taste panellists assessed the eating-quality of every animal on 100mm unstructured line scales, where 0= nil, 100= extreme for each eating quality attribute. Specifically, for juiciness 0= not juicy, 100= extremely juicy; for toughness 0= very tender, 100= very tough; for flavour 0= no lamb flavour, 100= full lamb flavour and for overall liking 0= least favourable impression, 100= most favourable impression. Carcass weight traits were also recorded. Further details of the phenotypic measurements have been provided in **Chapter 3**.

4.2.2 Genotyping and map construction

DNA was extracted from blood samples for all animals for a partial genome scan, covering chromosomes 1, 2, 3, 5, 14, 18, 20 and 21. These eight chromosomes were chosen because previous work had revealed QTL affecting meat and carcass traits on these chromosomes in sheep or in syntenic regions in other species (Kmiec, 1999; Broad *et al.*, 2000; Stone *et al.*, 1999; Freking *et al.*, 1999; Freking *et al.*, 2002; Nicoll *et al.*, 1998; Elo *et al.*, 1999; Walling *et al.*, 1998). Informative marker panels were developed separately for each sire, containing an average of 16, 8, 10, 6, 7, 6, 7, 4 informative microsatellite markers per sire on chromosomes 1, 2, 3, 5, 14, 18, 20 and 21, respectively. This was achieved by initially genotyping each sire for all available microsatellite markers across each candidate region and then selecting heterozygous markers at approximately 10 centiMorgan (cM) intervals wherever possible. All offspring were subsequently genotyped for selected markers that were heterozygous in their sire. In total, 139 markers were included in this study.

Marker locations were estimated by producing linkage maps using CriMap version 2.4 (Green *et al.*, 1990). These were in close agreement with previous studies (Maddox *et al.*, 2001), indicating accurate genotype data, and only differed for closely linked markers. In cases where there was a disagreement with regard to marker order between the published linkage maps, the marker order was checked using the CriMap-flips option. The marker order with the highest likelihood was chosen, in order to create a consensus linkage map that was used in subsequent QTL analyses.

4.2.3 QTL analysis

4.2.3.1 Treatment of data

Prior to QTL analyses the phenotypic data were subjected to a variance component analysis (**Chapter 3**) and significant fixed effects were identified. The fixed effects included in this analysis were: lamb sex (male or female), dam age (four classes: 2, 3, 4 and 5 years old), year of lamb birth (three classes: 2001, 2002 and 2003), management group (1 or 2), and litter size at birth (two classes: 1 or 2). Triplets comprised proportionately less than 0.02 of the data and were combined with the twin lambs in the classification of litter size. The model for meat quality traits included dam age (four classes), litter size (two classes) and date by year of slaughter (15 classes). The only interaction found to be significant was between fixed effects year of lamb birth and management group. Date of birth was fitted as a covariate for the CT traits.

The QTL analyses (described below) fitted these fixed effects simultaneously with estimates of the QTL effects. QTL analyses were performed for all meat quality traits, but a rationalised set of traits was chosen for the CT measures (only areas and densities of tissues at the three sites). Furthermore, genetic correlations had been previously calculated between all equivalent CT measures taken at different sites (**Chapter 3**); when the genetic correlation between the same measurement at different sites was greater than 0.8, the measurements were averaged after scaling by their standard deviations (otherwise they were treated as separate traits). Thus, measurements were averaged across sites for all traits except muscle density ISC, bone density ISC, bone area LV5, bone area TV8.

4.2.3.2 Information content

Information content was calculated at 1-cM intervals across all the regions under investigation in this study for each analysis. The information content of an individual marker is the proportion of animals in which the allele inherited from the sire can be unambiguously identified. Information content at genome position *i* was calculated as $Var(p_i)/0.25$ where p_i is the inheritance probability for each offspring included in the analysis and 0.25 is the expected variance of inheritance probabilities for a fully informative marker (Knott *et al.*, 1998).

4.2.3.3   Regression model

The genotyped pedigree contained nine half-sib families with 600 or 300 progeny, depending upon the trait. The analysis used the multimarker approach for interval mapping in half-sib families as described by Knott *et al.* (1996), as implemented by the web-based software package QTL Express (Seaton *et al.*, 2002), and as applied to QTL mapping studies in cattle by Spelman *et al.* (1996) and Vilkki *et al.* (1997). The method contains the following steps: for each offspring the probability of inheriting a particular sire haplotype is calculated at 1-cM intervals conditional on the linkage phase of the sire and marker genotypes of the individual and its sire. For a given position the conditional probabilities of the offspring provide independent variables on which the trait score can be regressed. Therefore, the regression model, in which fixed effects were simultaneously fitted, may be represented as:

$$y_{ij} = a_i + b_{ik} p_{ijk} + e_{ij} \quad (1)$$

where y_{ij} denotes the trait score of the jth individual originating from sire *i*; a_i is the average effect for half-sib family *i*; b_{ik} is the effect of one of the paternal haplotypes at interval *k* within half-sib family *i*; p_{ijk} is the probability for individual *j* of inheriting the first paternal haplotype at interval *k* conditional on the marker genotypes; and e_{ij} is the residual effect for individual *j*. Fixed effects fitted in these analyses were those found to be significant, for each trait, by the variance analysis.

For each regression, an *F*-ratio of the full model including the inheritance probabilities *vs.* the same model without the inheritance probabilities was calculated. In a one-QTL model, the location with the largest *F*-ratio was taken to be the best estimated position for a QTL for each trait.

4.2.3.4 Size of QTL effects

From the half-sib analysis the within-sire substitution effects were obtained for each sire. The average substitution effect was calculated for those sires that showed significant evidence of a segregating QTL, *i.e.*, for which the absolute value of the sire-specific *t*-statistic was significant ($P<0.05$). Results were expressed in residual standard deviation units (RSD).

For single-QTL analyses, the proportion of the phenotypic variance explained by the QTL, under the regression approach, may be obtained by $4 * (1- m.s._{full} / m.s._{reduced})$ (Knott *et al.*, 1996), where, $m.s._{full}$ is the model with one QTL fitted and $m.s._{reduced}$ is the model with no QTL fitted.

To address possible concerns of overestimating the size of QTL effects, QTL analyses were rerun by fitting background genetic effects on other linkage groups, *i.e.* cofactors, as suggested by Zeng (1993) and Jansen (1993). Background genetic effects were included as cofactors stepwise, beginning with the locus showing the largest estimated effect, followed by the locus with the second largest effect, until no further QTL were detected at the suggestive level of significance.

4.2.3.5 Significance Thresholds

Two significance thresholds were applied. The first level was the chromosome wise threshold, which takes account of multiple tests on a specific chromosome but does not correct for testing on the entire genome. Although calculated as an *F*-ratio, the distribution of the test statistic under the H_o of no QTL is unknown for half-sib analyses (De Koning *et al.*, 2001). Therefore, chromosome-wise significance thresholds were determined empirically by permutation for individual chromosome (Churchill and Doerge, 1994). One thousand permutations were studied for each trait and the relevant fixed effects and covariates were fitted. The chromosome-wise levels vary between chromosomes depending on their length and the markers they contain. Secondly, the genome-wise significance level (where, by chance, we expect 0.05 significant results per genome analysis) was obtained using the Bonferroni correction: $p_{genome\text{-}wise} = 1 - (1 - p_{chromosome\text{-}wise})^n$ (Knott *et al.*, 1998). For example, assuming 27 chromosomes are being analysed (*i.e.* there are 27 independent tests), the chromosomal test significance level would be p=0.001898 to give the genome-wise 0.05 level $((1-0.001898)^{27} = 1-0.05)$. None of these significance levels take the testing of multiple traits in the present and future studies into account.

For the genome-wise threshold levels, little variation was found in equivalent significance thresholds across different traits and across different chromosomes, and the appropriate *F*- value was invariably close to 3.0. Therefore a rounded *F*-value of 3.0 was used for 5% genome-wise significance.

4.2.3.6 Confidence Intervals

If the largest *F*-ratio indicated a QTL at the genome wise level, one- and two-LOD support intervals were produced by taking the region of the chromosome encompassed when reducing the largest *F*-ratio by the equivalent of a LOD score of either 1.0 or 2.0, to get 95% and 99% support intervals (Lander and Botstein, 1989). For comparison, bootstrap confidence intervals were also calculated (Visscher *et al.*, 1996).

4.2.3.7 Two QTL model

Where evidence was found for a single QTL on a chromosome at the genome-wise significant level, a model with two linked QTL on that chromosome was tested. This model was implemented by a grid search at 1 cM resolution of all possible positions for two QTL, the two chosen QTL positions being those that maximised the joint *F*-value testing the model of two QTL *vs.* no QTL. The significance of the second QTL was judged by deriving the *F*-value for the comparison of the best two QTL model *vs.* the best single QTL model for that linkage group. This *F*-value was tested against significance thresholds derived for the tests of one QTL *vs.* no QTL, as has previously been found to be appropriate (De Koning *et al.*, 2001).

4.3 Results

This study was successful in detecting QTL significant at both chromosome-wise and genome-wise level in seven out of eight chromosomal regions. All families produced evidence for significant QTL in one or more regions and 20 QTL were declared significant at the chromosome-wise level or greater. Two-QTL model analyses, performed on the nine

QTL that reached the 5% genome-wise significant level, never gave a statistically significantly better fit than the corresponding one-QTL models.

Descriptive statistics for the traits for which significant QTL were found are presented in Table 4.2. No significant QTL were identified for traits associated with direct measures of fatness (*i.e.* fat area, fat density, fat classification score, conformation score and subcutaneous fat) and the sensory traits of juiciness, toughness and overall liking.

Table 4.2 *Mean values, standard deviation (SD) and residual standard deviation (RSD) for traits showing significant QTL*

Trait[+]	Mean	SD	RSD[‡]
CT live weight (kg)	32.2	4.97	4.46
Average muscle area (mm²)	24.2	7.69	7.57
Muscle density ISC	43.3	2.85	1.59
Muscle density LV5-TV8	21.7	7.16	7.13
Bone area LV5 (mm²)	716	128	88.7
Bone area TV8 (mm²)	2846	527	359
Bone density ISC	317	39.3	25.3
Slaughter live weight (kg)	26.0	19.1	5.20
Hot carcass weight (kg)	18.0	3.22	2.55
Cold carcass weight (kg)	17.5	3.14	2.51
Colour a* (redness) (units)	17.8	1.88	1.37
Colour L* (lightness) (units)	40.5	3.24	2.08
Colour b* (yellowness) (units)	7.65	1.32	1.30
Lamb flavour (0-100)	26.5	6.39	4.71

[+]LV5=5th lumbar vertebrae, TV8= 8th thoracic vertebrae
[‡]Standard deviation of the phenotypic residual values after correcting for fixed effects and covariates included in the model.

A summary of the genome- (shown graphically in **Appendix I**) and chromosome-wise significant QTL is presented in Table 4.3. The size of effects and the proportion of variance attributable to the QTL are presented in Table 4.4. Quantitative trait loci reaching significance at the genome-wise levels were observed for a range of traits, including lamb flavour, muscle densities, live weight-related traits, and colour traits. The result supported by the strongest statistical evidence was a QTL affecting lamb flavour on chromosome 1

(Figure 4.1), for which the genome-wise significance level (*P*= 0.002) was obtained. The most likely position of this QTL is near marker *MAF64*. The QTL effect for lamb flavour was on average 1.89 residual standard deviation (RSD) in families S1, S3, S4 and S6. The proportion of the phenotypic variance explained by the QTL for lamb flavour was 0.44.

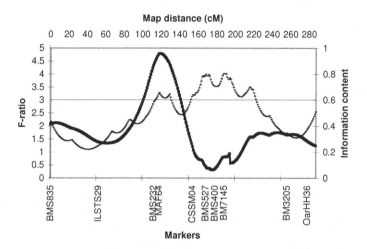

Figure 4.1 *F*-ratio profile for the QTL on sheep chromosome 1 affecting lamb flavour (■), and information content (+). Marker positions are indicated on the lower X-axis and map distances in cM are shown on the upper X-axis. The thin horizontal line indicates the 5% genome-wise significance threshold

A highly significant QTL affecting muscle density (LV5-TV8) (*P*= 0.0003) was identified on chromosome 2 with the estimated position at 28-cM lying between markers *CSSM47* and *FCB226* (shown in **Appendix I**). This QTL had an average effect of 1.51 RSD, in segregating families, and it accounted for 0.15 of the phenotypic variance in this trait (Table 4.4). Additionally, a QTL was identified on chromosome 3 for muscle colour a* (redness) (*P*= 0.005) (shown in **Appendix I**). The QTL was located at 113-cM, the size of the effect was 2 RSD and the proportion of the phenotypic variance due to the QTL was 0.28. A QTL for slaughter live weight was also identified on chromosome 1 (shown in **Appendix I**) and the proportion of the phenotypic variance explained by this QTL was 0.36. On chromosome 5, a QTL was found for cold carcass weight (*P*= 0.007) and hot carcass weight (*P*= 0.004) (shown in **Appendix I**) explaining 0.27 and 0.25 respectively, of the phenotypic variance, with an effect of 1.22 and 1.20 RSD at 0-cM, in the region of marker *TGLA176*. A QTL affecting bone density (ISC) was located on chromosome 1 close to the transferrin gene, with size effect of 1.14 RSD (shown in **Appendix I**). The QTL explained 0.13 of the phenotypic variance. Muscle density (ISC) produced evidence for a QTL at the 5% genome-wise level, on chromosome 3 and the proportion of the phenotypic variance explained by this QTL was 0.13 (shown in **Appendix I**). Finally, a QTL for slaughter live weight was, also, found on chromosome 2 (*P*= 0.003) at 262cM, explaining 0.33 of the phenotypic variance.

Table 4.3 *Summary of significant QTL, from across families analyses, exceeding the 5% genome- (F>3.0, in bold) and chromosome-wise significance level, presented by chromosome for carcass traits assessed by CT and meat quality traits*

Trait[+]	Chr[‡]	Position, cM	Marker Interval	F-ratio	5% Chr-Wise Threshold	1% Chr-Wise Threshold
Lamb flavour (0-100)	1	119	MAF64	**4.80**	3.15	3.91
Slaughter live weight (kg)	1	229	LCV105-BMS1789	**3.23**	2.98	3.66
Bone density ISC	1	261	BM3205-OarHH36	**3.15**	2.70	3.26
Hot carcass weight (kg)	1	227	LCV105-BMS1789	2.97	2.86	3.44
Colour b* (yellowness) (units)	1	165	INRA11-BMS527	2.55	2.55	2.95
Muscle density LV5-TV8	2	28	CSSM47-FCB226	**3.45**	3.45	3.95
Slaughter live weight (kg)	2	262	BM6444-BMS356	**3.02**	2.88	3.49
Colour a* (redness) (units)	3	113	KD0103-BL4	**3.31**	2.68	3.26
Muscle density ISC	3	172	BM6433-BMS772	**3.16**	2.60	3.15
Cold carcass weight (kg)	5	0	TGLA176	**3.23**	2.56	3.25
Hot carcass weight (kg)	5	0	TGLA176	**3.07**	2.72	3.27
Average muscle area (mm²)	5	116	MCM527-CSRD2134	2.44	2.88	3.27
Colour L* (lightness) (units)	18	80	ILSTS54-MCMA26	2.74	2.24	2.75
Bone area LV5 (mm²)	18	83	OB2-CSSM018	2.26	2.21	2.60
Colour L* (lightness) (units)	20	42	BM1815-DRB1	2.94	2.43	2.94
Bone area TV8 (mm²)	20	55	OMHC1	2.90	2.50	3.08
Bone density ISC	20	52	OLARDB-OMHC1	2.46	2.47	3.00
Bone area LV5 (mm²)	20	21	MCMA36-CP73	2.45	2.46	3.24
Hot carcass weight (kg)	21	88	HH22-BMC1948	2.72	2.48	3.14
CT live weight (kg)	21	11	BMC2228-BMC1206	2.41	2.08	2.76

[+]LV5=5th lumbar vertebrae, TV8= 8th thoracic vertebrae
[‡]Chr= chromosome

A further 11 QTL achieved significance at the 5% chromosome-wise level. These were for hot carcass weight (OAR 1 and OAR 21), muscle colour L* (lightness) (OAR 20 and OAR 18), bone area (TV8) (OAR 20), muscle colour b* (yellowness) (OAR 1), bone density (ISC) (OAR 20), bone area (LV5) (OAR 20 and OAR 18), muscle area (OAR 5), and live weight at CT scanning (OAR 21). The size of the effects for the chromosome-wise level significant QTL varied from approximately 0.6 to 1.8 RSD. The largest effect of 1.79 RSD was for hot carcass weight on chromosome 21. The four QTL that mapped to chromosome 20 were all located in the major histocompatibility complex (MHC) region.

Table 4.4 *Summary of phenotypic variance explained by the significant QTL for CT traits and meat quality traits, presented by chromosome*

Trait[+]	Chromosome	Families significant	Effect[‡] ±SE	Proportion of phenotypic variance due to QTL
Lamb flavour (0-100)	1	S1, S3, S4, S6	1.89±0.65	0.44
Slaughter live weight (kg)	1	S3, S4, S8	1.29±0.28	0.36
Bone density ISC	1	S2, S3, S5	1.14±0.26	0.13
Hot carcass weight (kg)	1	S3, S4	1.53±0.34	0.23
Colour b* (yellowness) (units)	1	S2	1.37±0.34	0.19
Muscle density LV5-TV8	2	S3, S4	1.51±0.28	0.15
Slaughter live weight (kg)	2	S2, S3, S4, S8	1.19±0.28	0.33
Colour a* (redness) (units)	3	S1	1.91±0.58	0.28
Muscle density ISC	3	S1, S6	1.51±0.41	0.13
Cold carcass weight (kg)	5	S3, S4, S5	1.22±0.27	0.27
Hot carcass weight (kg)	5	S3, S4, S5	1.20±0.27	0.25
Average muscle area (mm²)	5	S3, S4	1.35±0.29	0.09
Colour L* (lightness) (units)	18	S2, S4, S9	0.93±0.21	0.21
Bone area LV5 (mm²)	18	S3, S8	0.81±0.22	0.08
Colour L* (lightness) (units)	20	S5, S7	1.02±0.26	0.24
Bone area TV8 (mm²)	20	S5	0.57±0.21	0.12
Bone density ISC	20	S1, S4, S9	0.68±0.17	0.09
Bone area LV5 (mm²)	20	S2, S9	0.64±0.16	0.09
Hot carcass weight (kg)	21	S1, S3, S5	1.79±0.41	0.21
CT live weight (kg)	21	S5, S9	0.85±0.24	0.09

[+]LV5=5th lumbar vertebrae, TV8= 8th thoracic vertebrae
[‡]Effect is measured in residual standard deviation units and is averaged across families for which the QTL effect was significant

The analyses in which background genetic effects were fitted as cofactors generally did not alter the results because in most cases where multiple QTL were observed for the same trait, these QTL segregated in different families. The exceptions were slaughter weight and hot carcass weight, and for these two traits the apparent overestimation of the size of the QTL effects was reduced somewhat. For slaughter live weight, where the QTL on OAR 1 and OAR 2 segregated in the same families, the proportion of the phenotypic variance explained by the QTL was reduced to 0.31 for OAR 1 and to 0.21 for OAR 2. For hot carcass weight, where three QTL found on OAR 1, OAR 5 and OAR 21, with the same families segregating, the proportion of the phenotypic variance accounted for the QTL was 0.14 for OAR 1, 0.21 for OAR 5 and 0.13 for OAR 21, accounting for a total of 0.49 of the phenotypic variation.

Confidence intervals for the QTL significant at the genome-wise level are reported in Table 4.5. The one- and two-LOD support intervals were on average 17 and 41 cM long, respectively. Smaller LOD support intervals were typically obtained when the maximum F-ratio was high ($F>3.0$). The chromosome-wise significant QTL for lamb flavour, muscle density, bone density, colour a* (redness) and slaughter live weight all had relatively small tight confidence intervals as assessed by the LOD-drop method. As expected, the bootstrap confidence intervals are wider, and generally cover the greater part of the chromosome.

Table 4.5 *Confidence intervals for QTL significant at the 5% genome-wise significant level, presented by chromosome*

Trait[†]	Chromosome	1 LOD Support (95% CI)	2 LOD CI (99% CI)	Position, (cM)	Bootstrap C.I.
Lamb flavour (0-100)	1	111-131	106-136	119	5-269
Bone density ISC	1	246-279	234-287	261	0-287
Slaughter live weight (kg)	1	218-239	209-287	229	117-287
Muscle density LV5-TV8	2	17-40	7-49	28	6-227
Slaughter live weight (kg)	2	223-244	210-297	262	0-297
Colour a* (redness) (units)	3	106-117	93-125	113	63-205
Muscle density ISC	3	158-195	149-205	172	83-197
Cold carcass weight (kg)	5	0-15	0-21	0	0-133
Hot carcass weight (kg)	5	0-15	0-21	0	0-139

[†]LV5=5th lumbar vertebrae, TV8= 8th thoracic vertebrae

4.4 Discussion

This study has identified 20 significant QTL for a range of meat quality and carcass traits. Of these, nine were genome-wise and 11 chromosome-wise significant. This is novel and useful information that may provide opportunities for genetic improvement of meat and carcass quality. Traits for which QTL have been found include various definitions of carcass (hot and cold) and live weight (assessed by CT), lamb flavour, meat colour (a*, b* and L*), muscle density (ISC, LV5-TV8), muscle area, bone area (LV5 and TV8) and bone density (ISC). The last trait, bone density, may be of particular importance as an animal model for osteoporosis.

Chromosome 1 was chosen because of the presence of the transferrin gene at 272cM, which has been shown to be associated with growth (Kmiec, 1999). Five QTL were found to be segregating on chromosome 1 in this Blackface sheep population. The regression analysis revealed three significant QTL, at the genome level, for lamb flavour, slaughter live weight, and bone density (ISC) and two at the chromosomal level for hot carcass weight and colour b* (yellowness). Regarding the QTL for lamb flavour, there is no comparable published information available on sheep or other species, except for studies in pigs on pork flavour and 'boar taint' traits (androstenone, indole and skatole) (Quintanilla *et al.*, 2003; Lee *et al.*, 2004). Lee *et al.* (2004) identified QTL for boar flavour traits, as detected by sensory panel evaluations, in the same or adjacent marker intervals as QTL for chemically determined levels of androstenone, on pig chromosomes 6, 13 and 14. We used a comparative mapping approach to identify segments of conserved synteny between porcine chromosome 13 (SSC 13), human chromosome 21 (HSA 21), and the portion of ovine chromosome 1 (OAR 1) that contains the region for lamb flavour identified in this study. We found that a segment of HAS 21 shares conservation of synteny with SSC 13, indicated by the gene MX1 (Myxovirus resistance 1) which is flanked by markers *SW1056-S0215l* that are in the region in which Lee *et al.* (2004) identified a QTL for pork flavour, and it also shares conservation of synteny with OAR 1, indicated by *KRTAP6-1* (keratin associated protein 6-1). Therefore, the lamb flavour QTL on chromosome 1 is likely to be in the same relative position as the QTL for flavour in pigs described by Lee *et al.* (2004).

According to Young *et al.* (1997), whilst sheep meat odour/flavour is specifically caused by the medium branched chain fatty acids, there is a dietary interaction that intensifies the perception, and that is almost certainly caused by 3-methylindole (skatole). Skatole is associated with unpleasant odour in meat from entire male pigs (Claus *et al.*, 1994), and skatole (and indole) have long been known as microbial products of rumen bacteria (Yokoyama *et al.*, 1975; Hammond *et al.*, 1979). In the study of Young *et al.* (2003) it was concluded that skatole was the major cause of pastoral flavour in sheep meat, and that fat oxidation products represented a background flavour that varied with fatty acid profile. Young *et al.* (1999) suggest that the higher ratio of protein/non-fibrous carbohydrate characteristics of grass diets enhances skatole production through a higher deamination of protein amino acids by rumen microbes. Furthermore, in the study of Campbell *et al.* (2003), they have found a QTL for bone density in lean × fat backcross Coopworth sheep at the same location on OAR 1 as with the lamb flavour QTL identified in this study. This is of interest because we found the genetic correlation, estimated from the same dataset, between lamb flavour and bone density to be strongly negative (r_g= -0.67), although a biological explanation for this result is not obvious. In mice, Mehrabian *et al.* (2005) showed that QTLs for body fat and bone density both mapped to the vicinity of the *Alox5* locus on chromosome 6.

The QTL affecting slaughter live weight and hot carcass weight on chromosome 1 are in the same region as a QTL for live weight previously detected in Charollais sheep by McRae *et al.* (2005). The QTL for bone density was detected at 261cM in the region of the transferrin gene. Also, analyses of the phenotypes showed that the FAT line had significantly denser bone than the LEAN line (**Chapter 3**). This result was in contrast with the study by Campbell *et al.* (2003), who showed that bone density was generally negatively correlated with the amount of fat (subcutaneous, intramuscular, internal and total) (P< 0.01) in the carcass. Body weight and fat and muscle components are also correlated with bone density in humans. Low bone mineral density (BMD) has been shown to be an important factor in osteoporotic fracture risk in humans (Cummings *et al.*, 1990). Additionally, a potential QTL for cold carcass weight on chromosome 1, which had a nominal significance of P= 0.008, mapped to the same location as a QTL for slaughter live weight and segregated in the same families. Hence, it might be assumed that this is the same QTL expressing itself in different, but closely related, traits.

A QTL on chromosome 2 for muscle density (LV5-TV8) was detected near marker *CSSM47*. Muscle density is genetically correlated with intramuscular fat, and analysis of the FAT and LEAN lines revealed that the FAT line had a lower muscle density than the LEAN line (**Chapter 3**); this suggests more intramuscular fat, as fat is a less dense tissue than muscle. The QTL identified for muscle density (LV5-TV8) on chromosome 2 might also influence fat area and fat density, as QTL for these traits mapped to similar locations (26-cM and 30-cM, respectively) and segregated in the same families. However, the significance levels for these possible QTL were much lower, only reaching nominal significance levels of P= 0.0006 and P= 0.0005, respectively. But a QTL jointly affecting these traits is plausible: according to **Chapter 3**, fat area and fat density are strongly genetically correlated with muscle density. The evidence for a QTL on chromosome 2 affecting slaughter live weight was only significant at chromosome-wise level, and it was located 23cM distal to the GDF8 gene, responsible for the double muscling phenotype in cattle. Chromosome 2 was chosen for the mounting evidence of one or several QTL for carcass composition segregating around the GDF8 gene in sheep (Broad *et al.,* 2000 and Walling *et al.,* 2001). However, the region covered by the one LOD support interval for muscle density is approximately 200cM proximal from the region around GDF8 in which growth effects have been observed.

Two QTL were found on chromosome 3, one for each of muscle density (ISC) and colour a* (redness). The QTL affecting this muscle density and colour a* (redness) were located 55cM and 113cM, respectively, again not consistent with the candidate locus this chromosome was chosen for, insulin-like growth factor I gene (*IGF1*) which maps to 227cM. Other studies have identified QTL for carcass traits in homologous regions of chromosome 5 in cattle (Casas *et al.,* 2000; Davis *et al.,* 1998; Moody *et al.,* 1996). Also, associations with *IGF1* gene have been observed in other species. Collins *et al.* (1993), using markers closely linked to *IGF1*, detected a QTL associated with growth in mice. Casas-Carillo *et al.* (1997) also found a potential QTL associated with growth rate in pigs near *IGF1*. However, further studies will be needed to ascertain whether the same gene or genes is (are) responsible for the expression of the traits in different species. The detected QTL for colour a* (redness) is unique for sheep; however studies have been carried out in pigs. De Koning *et al.* (2001) found a suggestive QTL for colour a* (redness) on chromosome 13 (SSC 13) in pigs. We have previously shown that as lamb fatness increases the colour of meat becomes lighter (**Chapter 3**).

Three QTL, for hot carcass weight, cold carcass weight and for average muscle area, were observed on chromosome 5. The QTL for cold and hot carcass weight was located at 0cM, *i.e.* at the beginning of the mapped region of the chromosome, while the QTL for muscle areas was located at 116 cM, near the region the calpastatin gene maps to at 139cM.

Another significant region detected in the regression analyses was on chromosome 18. Chromosome 18 contains the Callipyge muscling gene (Cockett *et al.,* 1994), the rib-eye muscling locus (Nicoll *et al.,* 1998) and the Texel muscling QTL (Walling *et al.,* 2004). Two QTL were detected on that chromosome, one for colour L* (lightness) and one for bone area (LV5). Both these QTLs were located very close to the callipyge region, but biological inferences are difficult to draw for this result.

Four QTL for colour L* (lightness), bone areas (TV8 and LV5) and bone density, were identified on chromosome 20. The major histocompatibility complex (MHC) is located on chromosome 20, and contains genes coding for antigen presentation, that are critical to the acquired immune response. Studies in cattle (Elo *et al.*, 1999) and pigs (Walling *et al.,* 1998) have found effects for growth and fatness in homologous regions of their genomes. Biological links between acquired immunity and bone and meat attributes are currently not obvious to us.

4.5   Conclusions

In summary, this study has been successful in detecting QTL, that meet stringent significance thresholds, for a range of meat quality and carcass traits, including carcass and live weight, flavour of meat, meat colour, muscle density, muscle area, bone area and bone density. Bone density may have a lesser relevance to meat production but it may be of particular importance as an animal model for osteoporosis. These QTL offer several possibilities to breed sheep for improved meat quality. Verification of these results in independent populations is a logical progression. The results of these investigations coupled with previous variance component studies may enable us to design possible alternative breeding goals and selection strategies for meat quality traits.

APPENDIX I

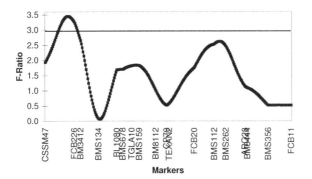

Figure 4.2 *F*-ratio profile for the QTL on sheep chromosome 2 affecting muscle density (LV5-TV8) (■). Marker positions are indicated on the lower X-axis. The thin horizontal line indicates the 5% genome-wise significance threshold

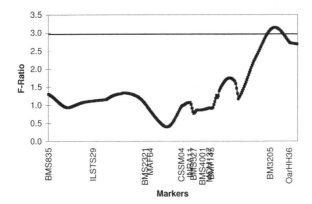

Figure 4.3 *F*-ratio profile for the QTL on sheep chromosome 1 affecting bone density (ISC) (■). Marker positions are indicated on the lower X-axis. The thin horizontal line indicates the 5% genome-wise significance threshold

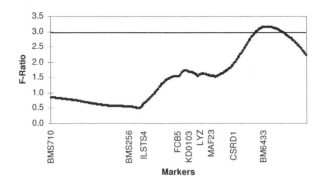

Figure 4.4 *F*-ratio profile for the QTL on sheep chromosome 3 affecting muscle density (ISC) (■). Marker positions are indicated on the lower X-axis. The thin horizontal line indicates the 5% genome-wise significance threshold

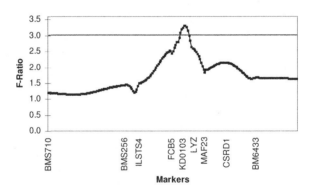

Figure 4.5 *F*-ratio profile for the QTL on sheep chromosome 3 affecting colour a[*] (■). Marker positions are indicated on the lower X-axis. The thin horizontal line indicates the 5% genome-wise significance threshold

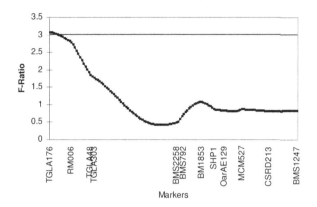

Figure 4.6 *F*-ratio profile for the QTL on sheep chromosome 5 affecting hot carcass weight (■). Marker positions are indicated on the lower X-axis. The thin horizontal line indicates the 5% genome-wise significance threshold

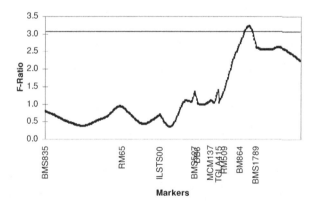

Figure 4.7 *F*-ratio profile for the QTL on sheep chromosome 1 affecting slaughter weight (■). Marker positions are indicated on the lower X-axis. The thin horizontal line indicates the 5% genome-wise significance threshold

Chapter Five

Genetic analyses and QTL detection, using a partial genome scan, for intramuscular fatty acid composition and eating quality

5.1 Introduction

There is increasing concern over the healthiness of the human diet, one recommendation being that the content of total fat be reduced, as it has an impact on human cardiovascular diseases, obesity and cancer (Simopoulos, 2001). In particular, lamb meat has been criticized because of high concentration of saturated fatty acids (SFA), presumed to increase the risk of heart disease, and a low polyunsaturated fatty acid (PUFA) to SFA ratio (Enser et al., 1996). However, it has become apparent that only some SFA have hypercholesterolaemic effects, and that certain monounsaturated fatty acids (MUFA) such as oleic acid (C18:1 c9) have beneficial effects (Schäfer, 2002). More recently, nutritionists have focused on the types of PUFA that lower blood cholesterol concentrations, and the balance in the diet between n-3 PUFA formed from linolenic acid (18:3) and n-6 PUFA formed from linoleic acid (18:2) (Williams, 2000).

Fatty acid composition is influenced by both genetic and environmental factors. Genetic factors have not been widely investigated in sheep, and it is not known whether fatty composition can be genetically manipulated towards a more favorable profile. In this thesis, we studied muscle fatty acid composition, because intramuscular fat cannot be easily removed, and, thus, inevitably has an impact on human health. The aims of this **Chapter** are, first, to investigate the inheritance of fatty acid composition in sheep meat; secondly, to estimate genetic relationships between fatty acids and computer tomograph-assessed muscle density (which previously established as being a visual predictor of intramuscular fatness); and thirdly, to identify quantitative trait loci for fatty acid composition. These results will jointly determine options for genetically improving the fatty acid profile of sheep meat.

5.2 Materials and Methods

5.2.1 Animals

Blackface lambs born at Roslin Institute's Blythbank farm, Scotland, during the period 2000 to 2003, were used as the experimental resource (more details in **Chapters 3 and**

4). The population comprised 300 lambs from 9 half-sib families for QTL detection. On average, families contained 33 male offspring for measurement of fatty acids traits, as described below, with a range from 12 to 46. Additionally, LEAN and FAT line lambs born in 2000 were also included in the study.

5.2.2 Phenotypic measurements

Phenotypic measurements of fatty acids were made on 350 8 month-old male grass-fed lambs, comprising 300 male lambs from the 9 backcross families, plus 25 LEAN and 25 FAT line male lambs born in 2000. Measurements were performed at the University of Bristol, on cohorts of 20 animals treated identically during their growth, transportation and pre-slaughter periods.

Lipids were extracted from 10g duplicate samples of *longissimus* (loin) muscle essentially as per Folch *et al.* (1957), separated into neutral and phospholipid, saponified and methylated, and individual fatty acids were separated by column chromatography and quantified as described by Demirel *et al.* (2004). Total fatty acids (used in this study) were taken as the sum of all of the phospholipid and neutral lipid fatty acids quantified. Total fatty acids included some unassigned fatty acids.

Fatty acid results are reported for major fatty acids and minor identifiable fatty acids relevant to the study. As a result of incomplete resolution, the *trans*-18:1 isomers are reported as a single value that does not include minor isomers (*trans*-13, *trans*-16-18:1) not resolved from *cis*-18:1*n*-9 and *cis*-18:1*n*-7. In addition to the minor cross-contamination of the latter two fatty acids, the fatty acid listed as *cis*-16: 1 consists of *n*-9 and *n*-7 isomers. Fatty acid data are presented as weight of fatty acids per weight of muscle tissue (mg/100 g).

Additional phenotypic measurements of carcass composition and meat quality were also taken on this population of lambs (**Chapter 3**). In particular, computerised tomography (CT) assessments of carcass composition were obtained on these lambs at 5 months of age, plus 350 females from the same families. From each scan image, the areas and image densities were obtained for the fat, muscle and bone components of the carcass. In **Chapter 3** we showed that muscle density was highly heritable and correlated with many meat quality traits influenced by intramuscular fat content. Thus, we assessed phenotypic and genetic correlations of muscle density with fatty acids.

5.2.3 Genotyping

The genotyping strategy was described in detail in **Chapter 4**. In summary, informative marker panels were developed separately for each sire, containing an average of 16, 8, 10, 6, 7, 6, 7, and 4 informative microsatellite markers per sire on chromosomes 1, 2, 3, 5, 14, 18, 20 and 21, respectively. Initially, each sire was genotyped for all available microsatellite markers across each candidate region, and heterozygous markers were then selected at approximately 10 centiMorgan (cM) intervals, wherever possible. All offspring were subsequently genotyped for selected markers that were heterozygous in their sire. In total, 139 markers were included in this study.

5.2.4 Data Analysis

All fatty acid measurement traits were right-skewed and were, therefore, log transformed prior to further analysis. Residual maximum likelihood (REML) methods were used to estimate variance components using an animal model, fitting the complete pedigree structure (4,847 animals), using ASReml (Gilmour *et al.,* 2004). The fixed effects included in this analysis were: line category (seven classes as given in **Chapter 3**) year of birth (2000, 2001, 2002 and 2003), management group (1 or 2), litter size (1= single or 2= twins and triplets) and year by slaughter day (15 classes). Triplets comprised proportionately less than 0.02 of the data, and hence were combined with the twin lambs in the litter size classification. The only significant interaction was between year and group.

It was investigated whether or not the primary data could be reduced to a small number of independent traits by a principal component analysis (PCA) (SAS, Proc Princomp, 2003) of each set of variables. Before PCA was carried out, the data were standardized by applying regression analysis to each trait, fitting fixed effects and recording the standardized residuals. The PCA was performed on these standardised residuals to create the new variables. Eigenvalues from this analysis indicated that four components provided a good summary of the 17 fatty acid measurements, accounting for 82% of the variance.

Heritability estimates were calculated for each log-transformed fatty acid, for total saturated fatty acids (SFA) (*i.e.,* the sum of the saturated fatty acids), for total monounsaturated fatty acids (MUFA) and for total polyunsaturated fatty acids (PUFA), using a bivariate analysis with CT-assessed muscle density. These analyses also yielded genetic and phenotypic correlations between each fatty acid measurement and muscle density.

Line Effects. To estimate genetic differences between the LEAN and FAT lines for each trait, using all of the data rather than just the 50 pure line lambs, true line effects were estimated as the generalized least squares solutions to equations describing the genetic composition of each line or cross. Details of the methodology for predicting means and variances for line categories have been provided elsewhere (**Chapter 3**). Line effects were estimated using untransformed fatty acid data.

5.2.5 QTL Analysis

QTL analysis was performed on the log-transformed data for each fatty acid, and for the three categories of total fatty acids (SFA, MUFA and PUFA).

Map Construction. Linkage analysis was carried out with the CriMap program, option 'build' (Green *et al.*, 1990). Marker locations were in close agreement with previous studies (Maddox *et al.*, 2001). In cases where there was a disagreement with the published linkage map, the marker order was checked using the CriMap-flips option. The marker order with the highest likelihood was chosen, in order to create a consensus linkage map that was used in subsequent QTL analyses.

QTL Detection. The method was described in **Chapter 4**.

Size of QTL Effects. For cases where QTL effects were significant, the within-sire substitution effects were obtained for each sire. The average substitution effect was calculated for those sires that showed significant evidence of a segregating QTL (*i.e.*, for which the absolute value of the sire-specific *t*-statistic was nominally significant ($P < 0.05$)).

For single-QTL analyses, the proportion of the phenotypic variance explained by the QTL ($\sigma^2_{QTL} / \sigma^2_P$) was calculated as $4^*(1\text{-Mean Square}_{full}/\text{Mean Square}_{reduced})$ (Knott *et al.*, 1996), where σ^2_{QTL} is the additive variance at the QTL.

Significance Threshold. Chromosome-wide empirical threshold values of the test statistics from the regression analysis, at $\alpha = 0.05$ chromosome level, were estimated with the permutation test procedure by Churchill and Doerge (1994). The thresholds vary between chromosomes depending on their length and the markers they contain.

Confidence Intervals. Bootstrap confidence intervals were estimated, as described by Visscher *et al.* (1996). Additionally, 95% support intervals for QTL location were obtained using the one-LOD drop approximation (Lander and Botstein, 1989).

Two-QTL Model. Once a single QTL on a chromosome had been identified, the presence of a second QTL was investigated by performing a grid search at 1-cM intervals. The two-QTL model was accepted if there was a significant improvement over the best possible one-QTL model at P< 0.05 using a variance ratio (*F*) test with 2 d.f. (for the additional additive effect and position estimated for the second QTL) as an approximate significance test.

5.3   Results

5.3.1 Summary Statistics

Summary statistics and estimated line differences for the fatty acid content of *longissimus* muscle are shown in Table 5.1. Concentrations of linoleic acid, arachidonic acid and DPA were lower in the LEAN line. The FAT line had a higher concentration of intramuscular fat (total fatty acids) than the LEAN line. Significant trends were also seen for myristic and stearic acid, with the FAT line having higher concentration of saturated fatty acids than did the LEAN line. The FAT line also had a higher content of oleic acid than the LEAN line, whereas conjugated linoleic acid (CLA) was at a lower concentration in the FAT line.

Table 5.1 *Line means (mg/100 g), trait phenotypic standard deviations, and line differences[1] (with standard error) for total fatty acids*

Trait	FAT Line	LEAN Line	Line difference (FAT-LEAN)	s.e. (Diff.)	s.d.
Saturated					
Myristic acid - 14:0	86.27	62.0	**24.3**	1.93	62.0
Palmitic acid - 16:0	515.1	509	6.1	4.44	515.1
Stearic acid - 18:0	405.4	389.1	**16.3**	2.20	405.4
Monounsaturated					
Palmitoleic acid – *cis* 16:1 (*n*-7, *n*-9)	54.78	55.92	-1.14	1.42	19.07
Oleic acid – *cis* 18:1 *n*-9	820.6	768.8	**51.8**	4.68	286
Cis-Vaccenic acid – *cis* 18:1 *n*-7	13.55	13.5	0.05	0.73	13.55
Vaccenic acid – *trans* 18:1 *n*-7	90.92	87.59	3.33	2.00	39.57
Gadoleic acid - 20:1	1.726	1.806	-0.08	0.25	0.654
Polyunsaturated					
Linoleic acid – *cis* 18:2 *n*-6	92.07	86.98	**5.09**	0.22	22.8
Linolenic acid – *cis* 18:3 *n*-3	39.47	40	-0.53	0.68	12.6
Dihomo-γ-linolenic acid - 20:3 *n*-6	3.894	3.518	0.376	0.27	0.76
Arachidonic acid - 20:4 *n*-6	34.36	32.46	**1.90**	0.85	6.95
EPA (Eicosapentanoic acid) - 20:5 *n*-3	29.93	28.14	1.79	1.11	6.69
Adrenic acid - 22:4 *n*-6	1.138	1.007	0.131	0.19	0.33
DPA (Docosapentaenoic acid) - 22:5 *n*-3	26.73	24.59	**2.14**	0.55	5.52
DHA (Docosahexaenoic acid) - 22:6 *n*-3	9.328	9.242	0.086	0.41	2.78
Conjugated fatty acid					
Conjugated linoleic acid (CLA) – 9-*cis*, 11-*trans* 18:2	32.89	36.67	**-3.78**	1.30	15.2
Total fatty acids (mg/100g)	2258	2150	**108**	7.80	795

[1] Line differences were estimated using the raw fatty acid data. Significant ($P< 0.05$) line differences are shown in bold.

5.3.2 Principal component analysis

The results of the principal component analysis are presented in Table 5.2 and graphically in Figure 5.1, for the four principal components (PC). The analysis showed that about 55% of the standardized variance is explained by the first principal component, 68.2% by the first two principal components, 76.5% by the first three principal components and 82% by the four principal components. Hence, 82% of the total variance in the 17 variables considered could be condensed into four new variables (PCs).

The first component (PC_1) showed approximately equal loadings on all input variables (Table 5.2) and hence may be interpreted as a descriptor of total fatty acid content. The second component (PC_2) explained 13.2% of the standardised variance. PC_2 had high positive loadings on long-chain (C_{20} and C_{22}) fatty acids and negative loadings on the rest of the fatty acids, thus expressing long-chain fatty acids versus saturated and monounsaturated fatty acids. The third component (PC_3) had high positive loading for the dihomo-γ-linolenic and adrenic acid (Table 5.2), and explained 8.29% of the variance in fatty acid measurements. Finally, the fourth component (PC_4) had a high positive loading on DPA but explained only 5.31% of the standardized variance.

Table 5.2 *Coefficients in the eigenvectors (loadings) for the first four principal components (PC) and their respective variances*

Variables	PC_1	PC_2	PC_3	PC_4
Myristic acid - 14:0	0.89	-0.12	-0.02	0.08
Palmitic acid - 16:0	0.95	-0.25	0.05	-0.02
Stearic acid - 18:0	0.92	-0.14	-0.01	0.02
Palmitoleic acid – *cis* 16:1 (*n*-7, *n*-9)	0.89	-0.21	0.05	-0.18
Oleic acid – *cis* 18:1 *n*-9	0.68	-0.45	0.29	-0.16
Cis-Vaccenic acid – *cis* 18:1 *n*-7	0.92	-0.23	0.01	-0.13
Vaccenic acid – *trans* 18:1 *n*-7	0.74	0.33	-0.21	-0.39
Gadoleic acid - 20:1	0.77	-0.16	0.33	0.15
Linoleic acid – *cis* 18:2 *n*-6	0.82	0.21	-0.32	0.20
Linolenic acid – *cis* 18:3 *n*-3	0.82	0.21	-0.32	0.20
Dihomo-γ-linolenic acid - 20:3 *n*-6	0.26	0.51	0.68	0.17
Arachidonic acid - 20:4 *n*-6	0.31	-0.54	0.44	0.34
EPA (Eicosapentanoic acid) - 20:5 *n*-3	0.18	0.77	-0.12	0.12
Adrenic acid - 22:4 *n*-6	0.16	0.47	0.60	-0.27
DPA (Docosapentaenoic acid) - 22:5 *n*-3	0.53	0.21	-0.22	0.56
DHA (Docosahexaenoic acid) - 22:6 *n*-3	0.22	0.62	0.27	-0.06
Conjugated linoleic acid (CLA) – 9-*cis*, 11-*trans* 18:2	0.79	0.22	-0.24	-0.36
SFA (saturated)	0.96	-0.17	0.00	0.03
MUFA (monounsaturated)	0.98	-0.06	-0.06	-0.16
PUFA (polyunsaturated)	0.90	0.36	0.09	0.18
% of total variance	55.0	13.2	8.29	5.31
Cumulative % variance	55.0	68.2	76.5	82.0

The loading plot (Figure 5.1) shows the location of each of the 20 variables in the multivariate space of the first two principal component loading vectors. In the plot the direction and the length of the vector indicated how each variable contributed to the first two principal components. The first principal component is represented by the horizontal axis and has positive coefficients for all 20 fatty acids, corresponding to the 20 vectors directed into the right half of the plot. The second principal component, represented by the vertical axis, has positive coefficients for total PUFA and negative for total SFA and total MUFA. This result indicates that this component distinguishes between meats that have high concentration for total PUFA and low for total SFA and MUFA, and vice versa.

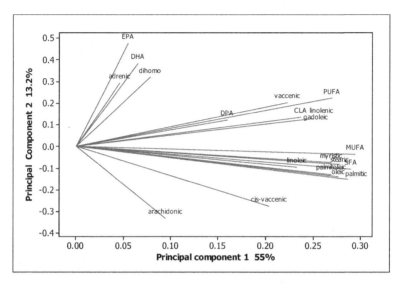

Figure 5.1 The dimensionally reduced fatty acid data. All fatty acids are plotted on the first and second principal components. The first component is a measure of total fatty acid content and the second component may be interpreted as a measure of long-chain (C_{20}-C_{22}) fatty acids *vs.* SFA and MUFA

5.3.3 Genetic Parameters

Heritability estimates for fatty acids (log-transformed fatty acids) are shown in Table 5.3. Almost all of the fatty acids were low to moderate heritable, with the exception of *cis*-vaccenic, vaccenic, CLA, and arachidonic acid, which were highly heritable. Traits describing combinations of fatty acids (*i.e.*, total SFA, MUFA, and PUFA) were generally moderately to highly heritable.

Phenotypic and genetic correlations between average muscle density and fatty acids are also presented in Table 5.3. Phenotypic correlations were low, except for intramuscular fat content, which was strongly negatively correlated with muscle density. Genetic correlations tended to be stronger than the phenotypic correlations. In particular, muscle density was strongly negatively genetically correlated with vaccenic acid, CLA, intramuscular fat content, total MUFA and total PUFA. Genetic correlations with the proportions of total fatty acids that were SFA, MUFA and PUFA were -0.23, -0.54 and 0.39, respectively.

Table 5.3 *Bivariate heritabilities (h^2) for fatty acids (with standard errors (s.e.)), and phenotypic and genetic correlations (with standard errors (s.e.)) of total fatty acids with CT-assessed muscle density[1]*

Trait	h^2 (s.e.)	Phenotypic correlation (s.e.)	Genetic correlation (s.e.)
Saturated			
Myristic acid - 14:0	0.19 (0.14)	0.01 (0.06)	-0.09 (0.15)
Palmitic acid - 16:0	0.29 (0.17)	-0.07 (0.07)	-0.30 (0.10)
Stearic acid - 18:0	0.24 (0.15)	-0.01 (0.06)	-0.10 (0.16)
Monounsaturated			
Palmitoleic acid – *cis* 16:1 (*n*-7, *n*-9)	0.31 (0.18)	0.02 (0.07)	-0.17 (0.15)
Oleic acid – *cis* 18:1 *n*-9	0.27 (0.17)	-0.02 (0.07)	-0.30 (0.15)
Cis-Vaccenic acid – *cis* 18:1 *n*-7	0.67 (0.16)	-0.35 (0.06)	-0.47 (0.14)
Vaccenic acid – *trans* 18:1 *n*-7	0.49 (0.17)	-0.11 (0.07)	-0.50 (0.18)
Gadoleic acid - 20:1	0.30 (0.17)	0.01 (0.07)	-0.24 (0.15)
Polyunsaturated			
Linoleic acid – *cis* 18:2 *n*-6	0.10 (0.09)	-0.19 (0.06)	-0.45 (0.14)
Linolenic acid – *cis* 18:3 *n*-3	0.30 (0.02)	0.01 (0.07)	-0.24 (0.15)
Dihomo-γ-linolenic acid - 20:3 *n*-6	0.12 (0.10)	-0.05 (0.06)	-0.22 (0.13)
Arachidonic acid - 20:4 *n*-6	0.60 (0.17)	0.08 (0.07)	0.28 (0.18)
EPA (Eicosapentanoic acid) - 20:5 *n*-3	0.21 (0.13)	0.01 (0.06)	-0.35 (0.16)
Adrenic acid - 22:4 *n*-6	0.22 (0.13)	-0.09 (0.06)	0.12 (0.15)
DPA (Docosapentaenoic acid) - 22:5 *n*-3	0.13 (0.12)	0.12 (0.06)	0.12 (0.12)
DHA (Docosahexaenoic acid) - 22:6 *n*-3	0.16 (0.10)	-0.14 (0.06)	-0.47 (0.15)
Conjugated fatty acid			
Conjugated linoleic acid (CLA) – 9-*cis*, 11-*trans* 18:2	0.48 (0.16)	-0.15 (0.06)	-0.60 (0.17)
Total fatty acids (mg/100 g)	0.32 (0.09)	-0.57 (0.04)	-0.67 (0.14)
SFA (saturated)	0.90 (0.16)	-0.30 (0.06)	-0.45 (0.14)
MUFA (monounsaturated)	0.73 (0.18)	-0.35 (0.06)	-0.60 (0.13)
PUFA (polyunsaturated)	0.40 (0.16)	-0.26 (0.06)	-0.56 (0.17)

[1] Heritabilities, and phenotypic and genetic correlations were estimated using the transformed fatty acid data.

5.3.4 QTL Results

A total of 21 QTL, significant at chromosome-wide level, were detected in six out of eight chromosomal regions, for 14 out of 17 fatty acids, and for every category of fatty acid. All families produced evidence for significant QTL in one or more regions. A summary of the chromosome-wide significant QTL, along with the proportions of variance attributable to the QTL and the confidence intervals for QTL location are presented in Table 5.4. The profile for the QTL affecting linolenic acid is shown in Figure 5.2.

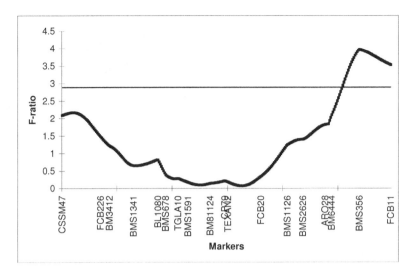

Figure 5.2 F-ratio profile for across-family QTL on sheep chromosome 2, affecting linolenic acid (18:3 n-3) (■). Marker positions are indicated on the lower X-axis. The thin horizontal line indicates the 5% chromosome-wide significance threshold

Results for fatty acids corrected for live weight were essentially identical to those for uncorrected fatty acids. Additionally, the two-QTL analyses never gave a significantly better fit than the single QTL analyses; hence, all results presented in Table 5.4 are from the single QTL analyses.

The QTL tended to be observed for individual fatty acids, rather than totals. In fact, the only QTL for combinations of fatty acids was for PUFA on chromosome 1. Bootstrapped confidence intervals were very large, typically covering the whole chromosome; however, it is known that this technique produces conservative intervals, around areas of higher marker density (Walling *et al.*, 1998b; 2002), and typically produces intervals covering the whole chromosome. The 1-LOD drop 95% support intervals were much smaller, typically 37 cM on average, ranging from 5 to 84 cM.

Table 5.4 *Most likely locations and effects of QTL detected for total fatty acids*

Trait	Chromo-some	Location (cM)[1]	5% threshold[2]	1 vs. 0 QTL	Additive (SE)	Families significant	% of phenotypic variance[4]
			F-ratios		Estimated effects[3]		
Saturated							
Myristic acid - 14:0	21	57 (6, 100)	3.02	3.14	3.03 (1.03)	S3, S7	26.1
Palmitic acid - 16:0	21	57 (0, 100)	2.94	3.17	3.03 (1.03)	S3, S7	29.0
Stearic acid - 18:0	21	58 (0, 85)	3.13	3.25	2.81 (1.00)	S3, S7	30.1
Monounsaturated							
Palmitoleic acid – *cis* 16:1 (*n*-7, *n*-9)	5	12 (0, 139)	2.63	2.77	1.07 (0.24)	S1, S5, S7	27.5
Oleic acid – *cis* 18:1 *n*-9	21	58 (0, 90)	2.98	3.23	2.80 (1.00)	S3, S7	27.2
Cis-Vaccenic acid – *cis* 18:1 *n*-7	21	58 (0, 85)	2.99	3.26	2.81 (1.00)	S3, S7	30.2
Gadoleic acid - 20:1	21	21 (0, 106)	3.03	3.49	2.39 (1.31)	S3, S7	33.1
	18	97 (23, 105)	2.01	2.38	0.82 (0.23)		18.4
Polyunsaturated							
Linoleic acid – *cis* 18:2 *n*-6	21	58 (0, 85)	3.14	3.22	2.80 (1.00)	S3, S7	29.7
Linolenic acid – *cis* 18:3 *n*-3	2	269 (6, 294)	2.89	3.97	1.66 (0.31)	S1, S8	37.6
	21	57 (0, 88)	3.08	3.17	3.03 (1.04)	S3, S7	29.1
Arachidonic acid - 20:4 *n*-6	21	58 (0, 72)	2.41	3.84	3.61 (0.32)	S5	20.3
	2	21 (0, 250)	2.73	2.75	1.62 (0.37)	S1, S3, S4	36.1
EPA (Eicosapentanoic acid) - 20:5 *n*-3	2	229 (6, 294)	2.70	3.05	1.34 (0.16)	S5, S8, S9	28.1
	1	79 (50, 280)	3.00	3.02	2.09 (0.45)	S1, S8	26.6
DPA (Docosapentaenoic acid) - 22:5 *n*-3	1	168 (50, 286)	3.19	3.52	1.63 (0.35)	S2, S6	32.7
	2	87 (6, 290)	3.32	3.49	1.60 (0.32)	S2, S3, S6	32.2
	21	0 (0, 107)	2.66	3.07	2.71 (0.65)	S6, S7	27.9
DHA (Docosahexaenoic acid) - 22:6 *n*-3	18	105 (25, 107)	2.31	2.32	0.94 (0.25)	S4, S8	24.2
Conjugated linoleic acid (CLA)	3	159 (83, 205)	2.51	3.22	1.50 (0.40)	S2	28.6

| PUFA | 1 | 85 (56, 286) | 2.59 | 2.84 | 1.87 (0.41) | S1, S8 | 23.9 |

[1] The 95% confidence intervals by bootstrap analysis are given in parentheses. The QTL were classified as significant at the 5% chromosome-wide level.
[2] The chromosome-specific 5% F value thresholds estimated via permutation analyses.
[3] The additive effects are given as the average effects within significant families.
[4] The proportion of phenotypic variance explained by the QTL is given as $\sigma^2_{QTL}/\sigma^2_p$

The most remarkable results were on chromosome 21 where 10 significant QTL were identified. Five of these were at position 58 cM, and a further three at position 57 cM. Moreover, apart from arachidonic acid, these QTL all segregated in the same families (S3 and S7). However, these individual fatty acids were not necessarily strongly correlated with each other. Residual correlations between the individual fatty acids were calculated as an approximation of phenotypic correlations. The correlations were seldom strong, ranging from 0.03 to 0.50.

5.4 Discussion

This study has produced novel information on the genetic control of fatty acid composition in sheep. The heritabilities indicate opportunities for genetically altering fatty acid composition, and the genetic correlations with muscle density suggest a means of genetic selection using *in vivo* measurements. Additionally, the study was successful in identifying 21 QTL for a range of fatty acids. The QTL tended to be for the individual fatty acids rather than for trait combinations, and QTL were found for each category of fatty acid (*i.e.*, SFA, MUFA and PUFA).

5.4.1 Line Differences

Selection for increased leanness changed some aspects of fatty acid composition, as the FAT line had a significantly higher intramuscular fat content (total fatty acids), and a higher content of stearic (18:0), oleic (18:1 *n-9*), linoleic (18:2 *n-6*) and DPA (22:5 *n-3*) fatty acids than the LEAN line. Stearic acid accounts for 18% of the total fatty acid in lamb meat (Rhee, 1992; Enser *et al.*, 1996). Several studies have shown that stearic acid is essentially neutral in its effects on serum total cholesterol, similar to oleic acid (Bonanome and Grundy, 1988; Zock and Katan, 1992; Kris-Etherton *et al.*, 1993; Grundy, 1994; Judd *et al.*, 2002). It is not clear why dietary stearic acid does not raise human serum cholesterol level as do other saturated fatty acids. Oleic acid is the primary monounsaturated fatty acid in lamb meat and accounts for about 32% of the total fatty acids (Rhee, 1992; Enser *et al.*, 1996). Available evidence indicates that, while the shorter chain saturated fatty acids raise human serum cholesterol concentrations, the monounsaturated oleic acid does not (Denke, 1994). The fatty acid composition of muscle influences important components of meat quality, such as meat flavor, flavor liking and overall acceptability. In the study of Cameron *et al.* (2000), where the effects of genotype, diet and their interaction on fatty acid composition of intramuscular fat in pigs were examined, it was found that linoleic acid was positively correlated with pork flavour (0.33), taste panel-assessed flavour liking (0.23), and overall acceptability (0.23), whereas DPA was negatively correlated with pork flavour (-0.30), flavour liking (-0.33) and overall acceptability (-0.30). In particular, linoleic acid (18:2 *n-6*), which is low in ruminant tissues (Wood *et al.*, 1999) (about 3.5% of total fatty acids), is a plant fatty acid that can be transformed to CLA (conjugate linolenic acid) by bacteria in the rumen (Kepler *et al.*, 1966). This low level of linoleic acid causes the polyunsaturated:saturated fatty acid ratio (an important nutritional index) to be below the recommended dietary value, which is 0.45 (Department of Health, 1994).

The LEAN line had more CLA than the FAT line, in agreement with the results of Wachira *et al.* (2002) in a comparison between (lean) Soay and (relatively fat) Suffolk breeds. Interest in CLA research has increased in the past few years as a result of reports of CLA consumption providing several health benefits (Kramer *et al.*, 1998). Because plants do not synthesize CLA, ruminant fats in meat are the primary dietary source of CLA for humans (Herbein *et al.*, 2000). It has been found that CLA has positive effects of reducing cardiovascular risk, protecting against atherosclerosis, is anti-carcinogenic, reduces intake, reduces body content of adipose tissue and lipid, and enhances the immune system (Cook *et al.*, 1993; Ip *et al.*, 1994; Lee et al, 1994; Nicolosi *et al.*, 1997; West *et al.*, 1998). In summary, these results show that, as the lamb becomes leaner, the polyunsaturated fatty acids of muscle increase and, in particular, quantities of the beneficial CLA increase.

5.4.2 Inheritance of Fatty Acids

The heritability estimates for most of the fatty acids measured in this study indicate that there is substantial genetic variation in these traits, such that fatty acid composition can potentially be improved, probably through indirect selection. In particular, saturated fatty acids were moderately heritable. The additive genetic coefficient of variation for these traits was 0.45 for myristic and 0.09 for palmitic and stearic acids, indicating that genetic alteration of these fatty acids is feasible. Due to lack of information on heritabilities of fatty acids in sheep, we will compare our results to published studies in other species. In pigs, heritability estimates for the content of palmitic and stearic fatty acids, obtained by Fernandez *et al.* (2003) for Iberian pigs, were 0.31 for palmitic and 0.41 for stearic acid. The estimate for palmitic acid is similar to the one obtained in the present study. The review by Sellier (1998) of three previous studies indicated that stearic acid had a higher heritability (0.51) than found in the present study.

The heritability estimates for monounsaturated fatty acids were high for *cis*-vaccenic (18:1 *n-7*) and vaccenic (*trans* 18:1 *n-7*) acid, and moderate for palmitoleic (16:1), gadoleic (20:1) and oleic (18:1 *n-9*) acid. In the study of Fernandez *et al.* (2003), the heritability of oleic acid was almost the same (0.30) as in our study.

The heritability of polyunsaturated fatty acids was variable with an average of 0.23. In pigs, Fernandez *et al.* (2003) and Sellier (1998) presented heritabilities for linoleic acid of 0.29 and 0.58, respectively. Linoleic acid appears to be more heritable in pigs than in sheep, although it should be noted that linoleic acid is much lower in ruminant tissues than in pigs, and in all species is obtained from the diet rather than being synthesized in the animal. The rumen biohydrogenates a high proportion of both linoleic and linolenic acid, reducing concentrations in body tissues (Wood *et al.*, 1994; 1999). Significant heritabilities for fatty acids not synthesized by the animal may be a result of different rates of fatty acid catabolism between animals.

Intramuscular fat (total fatty acids) was moderately heritable, although the value reported here is slightly lower than results from studies done in pigs. Malmfors and Nilsson (1979) obtained heritability estimates of 0.58 and 0.68 for Swedish Landrace and Large White pigs, respectively. Scheper (1979) reported a heritability estimate of 0.35 for German Landrace pigs, and a heritability of 0.55 was reported for Landrace pigs in Denmark (Just *et al.*, 1986) and in Switzerland (Schworer *et al.*, 1987). The average heritability weighted by number of sires for intramuscular fat content from previous reports is 0.53 (Sellier, 1998).

The heritabilities of trait combinations were surprisingly high in our dataset. One possible reason for this result is that, by combining correlated traits, the random measurement error or imprecision associated with individual fatty acids is reduced.

5.4.3 Relationships between Fatty Acids and Muscle Density

The results presented in **Chapter 3**, where we estimated genetic parameters for carcass composition, assessed in the live animal, and meat quality traits in these same sheep, demonstrated that altering carcass fatness will simultaneously change muscle density and aspects of intramuscular fat content. Hence, muscle density was chosen as the trait to be used for estimating phenotypic and genetic correlations with fatty acids, and we observed that muscle density was indeed strongly correlated with intramuscular fat content.

There is no published information on genetic relationships between muscle density, assessed by CT, and fatty acids in any species. Genetic correlations of saturated and monounsaturated fatty acids with muscle density were negative, with correlations involving vaccenic and *cis*-vaccenic acids being the strongest. Genetic correlations of polyunsaturated fatty acids with muscle density were mostly negative, except for arachidonic, adrenic, and DPA fatty acids. CLA, DHA, and linoleic fatty acids were strongly negatively correlated with muscle density. This result indicates that selection for decreased muscle density increases the concentration of CLA, DHA and linoleic fatty acids in lambs. In summary, selection to decrease muscle density, and, hence increase intramuscular fat, is expected to result in lamb meat with more total fatty acids. This is in accordance with the review of meat fatty acid composition by De Smet *et al.* (2004), where they reported that variation in fat content has an effect on fatty acid composition, independent of species or breed and dietary factors. Hence, the content of total SFA and total MUFA increases more rapidly with increasing fatness than does the content of total PUFA, leading to a decrease in the relative proportion of PUFA. This is shown at the phenotypic level in Figure 5.3.

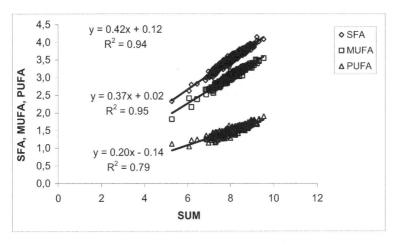

Figure 5.3 Relationships between total SFA, MUFA, and PUFA contents and their sum (mg per 100 g muscle) in intramuscular fat content

5.4.4 Identified QTL

Chromosome 1 was chosen for study because of the presence of the transferrin gene, which has been shown to be associated with growth effects (Kmiec, 1999), at 272 cM. Significant QTL on this chromosome affected two individual long-chain fatty acids, the EPA (20:5 *n-3*) and DPA (22:5 *n-3*) acids, and the total PUFA. The QTL for DPA is at the same relative position as the meat color trait yellowness (b[*]), which was reported in **Chapter 4** on the same Blackface population. Regarding QTL for fatty acids in sheep populations, there in no comparable published information, although there have been some studies in pigs. The importance of the long-chain *n-3* PUFA is that they are not incorporated into triacylglycerols to any important extent in ruminants, as they are in pigs (Wood *et al.*, 1999). They are restricted to phospholipids (mainly in membranes) and, therefore, are found in muscle but not fat tissue, except at very low levels in the total lipid (Enser *et al.*, 1996). This fact has important implications for PUFA supply in the diet of individuals consuming muscle and fat in the usual portions. Pig meat contributes about 80% of the total PUFA provided by pork, lamb, and beef, and about 60% of the *n-3* PUFA, 50% coming from pig fat (Enser *et al.*, 1996).

Chromosome 2 showed significant QTL for linolenic acid (18:3 *n-3*), DPA (22:5 *n-3*), EPA (20:5 *n-3*) and arachidonic acid (20:4 *n-6*). The QTL for linolenic acid mapped to the same position as slaughter live weight in these sheep (**Chapter 4**), and it was located 23 cM distal to the myostatin locus, which is responsible for the double muscling phenotype in cattle. Chromosome 2 was chosen for the mounting evidence of one or several QTL for carcass composition segregating around the myostatin locus (Broad *et al.*, 2000; Walling *et al.*, 2001; Johnson *et al.*, 2005a,b; Clop *et al.*, 2006). In the study of Clop *et al.* (2003), a significant QTL for linolenic fatty acid was identified on pig chromosome 12 (conservation of synteny with sheep chromosome 11). In the present study, the QTL on chromosome 2 affecting arachidonic fatty acid mapped to the same position as a QTL for muscle density in Blackface sheep (**Chapter 4**).

One QTL for CLA was found on chromosome 3, at 159 cM, excluding the candidate locus this chromosome was chosen for, insulin-like growth factor I, (*IGF-1*) at 227 cM. The QTL for muscle density was also significant and mapped to a position (172 cM) close to the CLA fatty acid content QTL. The concentration of CLA in lamb muscle is important in human nutrition, because it has been linked to a multitude of metabolic effects, including inhibition of carcinogenesis, reduced fat deposition, altered immune response, and reduced serum lipids (Mulvihill, 2001).

Another significant region was detected on chromosome 21. Remarkably, eight significant QTL (Figure 5.4) for arachidonic (20:4 *n*-6), *cis*-vaccenic (18:1 *n*-7), stearic (18:0), oleic (18:1 *n*-9), linoleic (18:2 *n*-6), palmitic (16:1), linolenic (18:3 *n*-3), and myristic (14:0) fatty acids were mapped to the same position (57-58 cM), and gadoleic (20:1) acid and DPA (22:5 *n*-3) were mapped to 21 cM and 0 cM, respectively. Additionally, the families segregating for eight out of 10 QTL identified, at the genome level, were the same. Hence, these effects likely correspond to a single QTL. Results from other studies in Iberian x Landrace pigs have identified QTL affecting linoleic fatty acid on pig chromosome 4 (Perez-Enciso *et al.*, 2000; Clop *et al.*, 2003) (conservation of synteny with sheep chromosome 25). Also, Perez-Enciso *et al.* (2000) have found that effects on backfat thickness, backfat weight, and *longissimus* muscle area in pigs were also significant and mapped to the same position as the linoleic acid content QTL. Linoleic acid is an essential fatty acid for mammals, because they lack desaturase capacity beyond the 9th carbon atom (Vance and Vance, 1996). It is a key component of cellular membranes and a precursor of prostaglandins and thromboxanes. It is also stored in adipose tissue or β-oxidized for energy production. In fact, it is highly digestible and preferentially deposited compared with other fatty acids (Lawrence and Fowler, 1997). High linoleic acid content is also associated with toughness and low consumer acceptability in pig meat (Whittington *et al.*, 1986; Cameron and Enser, 1991; Lawrence and Fowler, 1997).

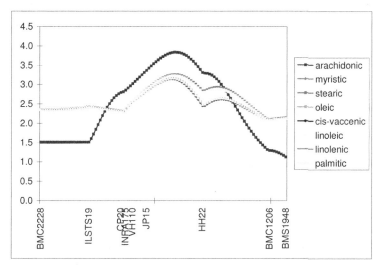

Figure 5.4 F-ratio profile for across-family QTL on sheep chromosome 21, affecting myristic acid (14:0), palmitic acid (16:0), stearic acid (18:0), oleic acid (cis 18:1 n-9), cis-vaccenic acid (cis 18:1 n-7), linoleic acid (cis 18:2 n-6), linolenic acid (cis 18:3 n-3), and arachidonic acid (20:4 n-6). Marker positions are indicated on the lower X-axis. The 5% chromosome-wide significance thresholds for each fatty acid are shown in Table 5.4

5.5 Conclusions

In summary, this study has shown that altering carcass fatness has simultaneously changed oleic acid (18:1 *n-9),* linoleic acid (18:2 *n*-6), DPA (22:5 *n*-3), CLA, and intramuscular fatness (total fatty acids). Heritabilities for most of the fatty acids were moderate to high, suggesting that composition of lamb meat can be changed genetically. Also, selection to alter muscle density and intramuscular fat content would also alter both the total quantities and proportions of saturated and monounsaturated fatty acids in lamb meat. Hence, the opportunity exists to use muscle density as a tool to breed pasture-finished lambs that have the potential to compete as a health-promoting food with other health oriented products on the market.

This study was also successful in detecting significant QTL related to fatty acid composition on sheep chromosomes 1, 2, 3, 5, 14, 18, and 21, and the same QTL were detected irrespective of whether the data were corrected for live weight or not. These QTL are clearly of potential importance to the sheep industry; however, they first need to be confirmed in independent populations, and more precise genetic markers would be required. Several genes might be selected as functional candidate genes to explain the QTL found in this study, as fatty acid metabolism can be influenced by a large number of genes involved in complex metabolic routes (Clop *et al.*, 2003). Association analyses between allelic variants of these genes and fatty acid content would need to be performed to find the necessary genetic markers.

5.6 Implications

This study has identified new information on the genetic basis of intramuscular fatty acid composition in sheep meat. We have demonstrated that quantities of different fatty acids are moderately to highly heritable; we have shown genetic correlations between live animal measurements and fatty acid composition; and we have found several QTL for fatty acid composition. These include eight QTL in one location on chromosome 21 with large effects on myristic, palmitic, stearic, oleic, *cis*-vaccenic, arachidonic, linoleic and linolenic acid content of intramuscular fat. Thus, we have demonstrated that it is possible, in principle, to breed for altered fatty acid composition, and we have gone a long way towards providing the tools for achieving this. The next step may involve finer mapping of the QTL detected, and analysis of these QTL in commercial populations.

Chapter Six

Genetic analyses of sensory characteristics and their relationship with fatty acid composition

6.1 Introduction

The most important aspect of meat quality is eating quality, usually defined as scores given by taste panellists for toughness, juiciness and flavour. These characteristics are affected by several factors in production, such as breed and diet (Wood et al., 2004a). It is known that muscles differ in the amount and fatty acid composition of the main lipid classes, neutral and phospholipids, these two constituting marbling fat. Variations in these fatty acids explain some of the quality differences between muscles, for example in shelf life and flavour. It is also known that variation in the amount and fatty acid composition of the lipids classes explain meat quality differences brought about by breed and diet, although this is more controversial (Wood et al., 2004b). Fat and fatty acids are important in their own right because of their effects on human health and it is important to select production options which maximise both meat quality and healthiness in meat production (Kouba et al., 2003).

Fat, especially animal fat, has been the subject of much interest and debate because of risks of some diseases when consumed in excess. Fat however is not only a concentrated source of energy for the body, the fat in meat also provides flavour, aroma and texture. When eaten, fat is also a carrier of the fat-soluble vitamins A, D, E and K and the essential fatty acids, and is important in growth and in the maintenance of many body functions (Nürnberg et al., 2005). As far as most consumers are concerned, meat should contain only a small amount of fat. However, some fat is always present in meat and indeed is required to impart flavour and juiciness (Melton, 1991). Many reports also show positive effects of fat level on tenderness (Wood, 1990).

Lipid oxidation in muscle systems is initiated at the membrane level in the phospholipid fractions as a free-radical autocatalytic chain mechanism (Labuza, 1971) in which prooxidants interact with unsaturated fatty acids resulting in the generation of free radicals and propagation of the oxidative chain (Ashgar et al., 1988). Also, lipid oxidation is a major cause of deterioration in meat quality (Gray and Pearson, 1987; Gray *et al.*, 1996). It limits the storage or shelf life of meat exposed to oxygen under conditions where microbial spoilage is prevented or reduced such as refrigeration or freezing. The products of fatty acids oxidation produce off-flavours and odours usually described as rancid (Gray and Pearson, 1994).

The dataset analysed in this thesis provides a unique opportunity to investigate relationships between fatty acid composition and meat eating quality assessments. Thus, this **Chapter** aims to examine, first, the inheritance of eating quality of muscle *longissimus thoracis et lumborum* (LTL) from grass-fed lambs; and secondly, to investigate relationships between eating quality assessments and fatty acid composition of meat.

6.2 Materials and Methods

6.2.1 Animals

The population structure is given in **Chapter 3**. Phenotypic measurements of eating quality traits and fatty acids were made on 350 8-month old male grass-fed lambs, *i.e.* 300 lambs from the QTL families, plus 25 LEAN and 25 FAT line male lambs born in 2000. Measurements were performed at the University of Bristol, on cohorts of 20 animals treated identically during their growth, transportation and pre-slaughter periods.

6.2.2 Sensory (taste panel) analysis

Descriptive sensory analyses were performed using a trained taste panel. A section of muscle *longissimus thoracis et lumborum* (LTL) was removed 24h after slaughter from the left side of the carcass, was packaged under vacuum and conditioned at $1^{\circ}C$ for a further 10 days when it was frozen at $-20^{\circ}C$ prior to assessment of eating quality under standardised conditions. Samples were thawed at $4^{\circ}C$ overnight and cut into 2.5cm chops, which were grilled to an internal temperature of $78^{\circ}C$, as measured by a thermocouple inserted into the centre of the muscle. Sensory descriptors were defined (Table 6.1) and ten experienced panellists rated the intensities of lamb flavour, abnormal, acidic, ammonia, bitter, fatty/greasy, fishy, grassy, livery, metallic, rancid, soapy, stale, sweet, vegetable, juiciness, and toughness as well as the hedonic overall liking on every animal on a 100mm unstructured line scale with anchor points at each end, where 0 meant no flavour or dislike extremely, and 100 meant very intense flavour or like extremely. The hedonic scale served as an indication of preference by the panel, but it cannot be used to infer consumer

111

acceptance since the results are based on ten assessors who can no longer be considered as typical consumers because of the training they have received in meat assessment.

Table 6.1 *Definition of eating quality descriptors*

Attributes	Definition
Lamb	Flavour associated with cooked lamb: *no lamb flavour to full lamb flavour*
Abnormal	Abnormal flavour not found in cooked lamb: *none to strong off-flavour*
Acidic	Sour taste
Ammonia	Flavour associated with ammonia
Bitter	Bitter taste
Fatty/greasy	Flavour associated with oil
Fishy	Flavour associated with fish
Grassy	Flavour associated with fresh grass
Livery	Flavour associated with liver
Metallic	Flavour associated with meat taste
Rancid	Rancid flavour found in the meat
Soapy	Flavour associated with soap
Stale	Flavour associated with stale meat
Sweet	Sweet taste
Vegetable	Flavour associated with vegetables
Juiciness	Amount of moisture released: *not juicy to extremely juicy*
Toughness	: *very tender to very tough*
Overall liking	Hedonic liking from the panellists

6.2.3 Lipid extraction and fatty acid composition

Lipids were extracted from 10g duplicate samples of *longissimus* (loin) muscle essentially as per Folch *et al.* (1957), separated into neutral and phospholipid, saponified, methylated, and individual fatty acids separated by column chromatography and quantified as described by Demirel *et al.* (2004). Total fatty acid was taken as the sum of all the phospholipid and neutral lipid fatty acids quantified. Total fatty acids included some unassigned fatty acids. Detailed descriptions of fatty acid extraction and analysis were given in **Chapter 5**.

Fatty acid data are presented as mg/100g of muscle. Two ratios of fatty acids types were calculated as indices of nutritional value. The polyunsaturated: saturated fatty acid ratio (PUFA: SFA) was defined as the sum of polyunsaturated to saturated fatty acids. The ratio of the *omega-6: omega-3* fatty acids was also calculated.

6.2.4 Data analysis

Residual maximum likelihood (REML) methods were used to estimate variance components using an animal model, fitting the complete pedigree structure (4847 animals), using ASReml (Gilmour *et al.*, 2004). The fixed effects included in the analysis of sensory traits were: line category (seven classes as given in **Chapter 3**), year of lamb birth (four classes: 2000, 2001, 2002 and 2003), management group (1 or 2), litter size (two classes: 1 or 2), year by slaughter day (15 classes) and panel number (26 classes). The model for fatty acids included the same fixed effects used for the sensory traits, apart from the panel number. Triplets comprised proportionately less than 0.02 of the data and were combined with the twin lambs in the classification of litter size. The only interaction found to be significant was between fixed effects year and group. Heritability estimates were then calculated for each eating quality trait, using an animal model, fitting the complete pedigree structure (4847 animals), using ASReml (Gilmour *et al.*, 2004).

Principal component analysis (PCA) (SAS, Proc Princomp, 2003) was performed on the primary data of eating quality traits alone, and eating quality traits and fatty acids. The PCA was used to reduce the primary data into a smaller number of independent traits, and also to investigate any relationships between eating quality traits and fatty acids composition. Before PCA were carried out, the data were standardised by applying regression analysis on each trait, fitting fixed effects and keeping the standardised residuals. The PCA were performed on these standardised residuals to create the new variables. Eigenvalues from this analysis indicated that five components provided a good summary of the 18 eating quality traits, accounting for the 80.8% of the variance. Also, six components accounted for 81.2% of the total variance for eating quality traits (juiciness, lamb flavour, abnormal flavour, toughness and overall liking) and fatty acids.

The dataset was not large enough to do bivariate genetic analyses with acceptable precision. Therefore, residual correlations between the individual eating quality traits with fatty acids were estimated using REML, as an approximation to phenotypic correlations. The model included the fixed effects of line category, year of lamb birth, management group, litter size, year by slaughter day and panel number. Sire was included as a random effect. The residual correlations were tested for significance (P<0.05).

Line Effects. In order to estimate genetic differences between the LEAN and FAT lines for each trait, using all the data rather than just the 50 pure line lambs, true line effects were estimated as the generalised least squares solutions to equations describing the genetic composition of each line or cross. Details of the methodology for predicting means and variances for line categories have been provided elsewhere (**Chapter 3**).

6.3 Results

6.3.1 Summary statistics

Summary statistics for eating quality traits are shown in Table 6.2, with significant (*P*<0.05) line differences shown in bold. The FAT line meat was perceived to be more juicy than the LEAN line meat. A significant trend was also seen for vegetable flavour, with the FAT line meat being more associated with vegetable flavour than the LEAN line meat. Non-significant trends, in the same direction, were also seen for abnormal flavour, bitter and metallic flavour. The LEAN line meat, although not significant, was perceived to have more normal 'lamb' flavour than the FAT line.

Table 6.2 *Predicted line means (0-100 scales), trait phenotypic standard deviations, and line differences (with standard error) for sensory panel traits. Significant (p<0.05) line differences are shown in bold*

Trait	FAT Line	LEAN Line	s.d.	Line difference (FAT-LEAN)	s.e. (Diff.)
Abnormal flavour	21.10	18.78	8.03	2.32	1.22
Acidic	4.826	3.791	3.19	1.03	0.79
Ammonia	1.166	1.106	1.91	0.06	0.65
Bitter	5.820	4.421	3.96	1.40	0.85
Fatty/greasy	16.52	16.99	5.48	-0.47	0.96
Fishy	1.283	1.107	1.49	0.17	0.57
Grassy	1.600	1.597	4.88	0.003	0.24
Juiciness	43.38	40.65	2.08	**2.73**	1.14
Lamb flavour	25.20	26.86	5.93	-1.66	1.04
Livery	12.58	12.15	7.19	0.43	1.17
Metallic	6.002	4.641	6.19	1.36	0.85
Overall liking	22.02	21.96	3.28	0.06	1.22
Rancid	3.528	3.225	3.04	0.30	0.79
Soapy	4.906	4.698	4.86	0.21	1.02
Stale	4.932	4.637	3.66	0.29	0.82
Sweet	6.300	7.192	3.68	-0.89	0.77
Toughness	36.32	36.37	9.98	-0.05	1.16
Vegetable	6.555	3.208	3.68	**3.35**	1.14

6.3.2 Principal component analysis

The results of the principal component analysis for the 18 eating quality traits are presented in Table 6.3 and graphically in Figure 6.1. The five first principal components explain 80.8% of the total variation (40.1, 18.6, 10.9, 6.10 and 5.10, respectively). Each PC represents an independent cause of variation, thus traits near each other are positively correlated, traits separated by 90° are independent and traits separated by 180° are negatively correlated. All PC are linear combinations of traits, but a trait far from origin that lies on a PC is predominant in defining this PC.

The first component (PC_1) explained 40.1% of the standardised variance and can be interpreted as a descriptor of abnormal meat flavours (Table 6.3). Figure 6.1 shows a plot of the traits on the first two principal components. Two groups were clearly distinguished lying on the first PC far from the origin. The first group included abnormal meat flavours (abnormal flavour, bitter, fatty/greasy, rancid, and stale). These variables are correlated and negatively correlated with the other group, the normal meat flavours (lamb flavour and overall liking), lying near the first PC on the opposite side. The second PC_2 grouped juiciness, grassy and sweet flavour, as an independent cause of variation. Toughness is not well presented by the first two PCs, lying near the origin. The third PC_3 explained 10.9% of the total variance and essentially represents toughness of meat. Toughness is negatively correlated with overall liking and juiciness (Table 6.3). The fourth PC_4 had high positive loadings on metallic and acidic flavours and explained 6.10% of the variance. Finally the fifth PC_5 was represented mostly by fishy flavour, but explained only 5.10% of the standardised variance.

Table 6.3 *Coefficients in the eigenvectors (loadings) for the first five principal components (PC) and their respective variances*

Trait	PC1	PC2	PC3	PC4	PC5
% of variance accounted for	40.1	18.6	10.9	6.10	5.10
Loadings:					
Abnormal flavour	**0.39**	0.00	-0.04	-0.05	-0.08
Acidic	**0.24**	0.03	-0.06	**0.46**	0.17
Ammonia	**0.20**	-0.25	-0.27	-0.15	-0.04
Bitter	**0.27**	0.08	0.16	0.09	-0.19
Fatty/greasy	**0.30**	0.18	-0.14	-0.02	0.24
Fishy	0.20	0.06	-0.10	-0.19	**0.59**
Grassy	0.16	**0.44**	-0.16	-0.11	-0.14
Juiciness	0.04	**0.37**	-0.34	**0.36**	-0.24
Lamb flavour	-0.33	0.07	-0.10	0.10	0.05
Livery	0.17	-0.28	-0.39	-0.11	-0.23
Metallic	0.16	-0.14	-0.06	**0.68**	0.21
Overall liking	-0.28	0.17	-0.31	0.00	**0.26**
Rancid	**0.30**	-0.14	-0.05	-0.09	0.05
Soapy	**0.24**	-0.26	-0.06	-0.03	-0.28
Stale	**0.29**	0.04	0.23	-0.07	0.02
Sweet	0.11	**0.54**	-0.09	-0.11	-0.26
Toughness	0.07	0.13	**0.63**	0.15	-0.09
Vegetable	0.19	0.22	0.11	-0.22	**0.34**

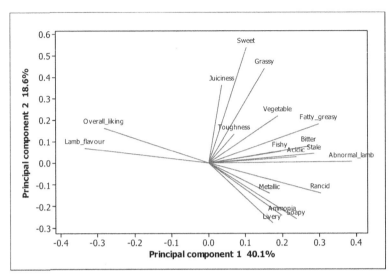

Figure 6.1 The dimensionally reduced taste panel data. The first component is a measure of abnormal meat flavours and the second component group comprises juiciness, grassy and sweet flavours

The results of the principal component analysis of the fatty acids and juiciness, lamb and abnormal flavour, toughness and overall liking are presented in Table 6.4 and graphically in Figure 6.2. The first six components explain 81.2% of the standardised total variation (41.4, 16.2, 7.4, 6.5, 5.9, and 3.8, respectively).

The first PC (PC_1) explained 41.4% of the total variability in fatty acid composition. This component was mainly characterised by total composition of fatty acids, while juiciness is not well represented by the first two PCs, lying near the origin. The second PC essentially distinguished two groups, lying far from the origin. The first group represents normal sensory variables (lamb flavour, overall liking) correlated with arachidonic acid (C20:4) (arachidonic produced from the conversion of linoleic acid by a series of desaturation and elongation steps) and the ratio of *n-6: n-3*. On the opposite side of the PC_2 the second group represents abnormal sensory variables (abnormal flavour, fatty/greasy, rancid, stale, bitter, vegetable, sweet, grass, acidic, ammonia, metallic, fishy and livery).

Table 6.4 *Coefficients in the eigenvectors (loadings) for the first six principal components (PC) and their respective variances*

Trait	PC1	PC2	PC3	PC4	PC5	PC6
% of variance accounted for	41.4	16.2	7.40	6.50	5.90	3.80
Loadings:						
Abnormal flavour	-0.06	**0.28**	0.15	**0.42**	-0.09	-0.16
Lamb flavour	0.07	-0.21	-0.19	**-0.47**	0.08	-0.04
Juiciness	0.00	-0.09	0.08	0.05	**-0.30**	**-0.57**
Toughness	-0.07	0.04	0.15	0.06	-0.09	**0.61**
Overall liking	0.08	-0.16	-0.25	**-0.41**	0.11	-0.29
Myristic acid - 14:0	**0.25**	-0.08	0.00	0.11	-0.22	-0.04
Palmitic acid - 16:0	**0.26**	-0.13	0.00	0.10	-0.03	0.00
Stearic acid - 18:0	**0.26**	-0.09	-0.01	0.07	-0.04	0.06
Palmitoleic acid – *cis* 16:1 (*n*-7, *n*-9)	**0.25**	-0.12	-0.02	0.11	0.06	-0.05
Oleic acid – *cis* 18:1 *n*-9	**0.26**	-0.13	-0.04	0.08	0.09	-0.01
Cis-Vaccenic acid – *cis* 18:1 *n*-7	0.18	-0.19	0.13	0.17	0.13	-0.04
Vaccenic acid – *trans* 18:1 *n*-7	0.21	0.14	-0.13	0.13	**0.27**	-0.03
Gadoleic acid - 20:1	**0.24**	0.09	-0.13	-0.01	-0.08	0.12
Linoleic acid – *cis* 18:2 *n*-6	0.21	-0.09	0.23	0.05	-0.08	-0.04
Linolenic acid – *cis* 18:3 *n*-3	**0.24**	0.09	-0.13	-0.01	-0.08	0.12
Dihomo-γ-linolenic acid - 20:3 *n*-6	0.08	0.16	**0.47**	-0.18	-0.01	-0.16
Arachidonic acid - 20:4 *n*-6	0.09	**-0.26**	0.25	-0.24	0.00	0.23
EPA (Eicosapentanoic acid) - 20:5 *n*-3	0.06	0.35	-0.06	-0.14	-0.03	0.02
Adrenic acid - 22:4 *n*-6	0.05	0.14	**0.40**	-0.05	0.19	-0.21
DPA (Docosapentaenoic acid) - 22:5 *n*-3	0.16	0.12	-0.13	-0.12	**-0.39**	0.05
DHA (Docosahexaenoic acid) - 22:6 *n*-3	0.07	**0.26**	0.15	-0.14	0.11	-0.09
Conjugated linoleic acid (CLA)	0.22	0.10	-0.14	0.16	0.21	-0.06
Total SFA[a]	**0.27**	-0.10	0.00	0.10	-0.12	0.00
Total MUFA[b]	**0.27**	-0.04	-0.05	0.12	0.12	0.00
Total PUFA[c]	**0.26**	0.14	0.08	-0.09	-0.04	0.02
PUFA: SFA[d]	0.07	**0.36**	0.14	-0.28	0.08	0.03
n-3[e]	0.21	**0.27**	-0.08	-0.14	-0.17	0.05
n-6[f]	**0.24**	-0.01	0.24	-0.18	0.00	0.11
n-6: n-3[g]	-0.01	**-0.35**	**0.34**	0.00	0.20	0.06

[a] Sum of saturated fatty acids: C14:0 + C16:0 + C18:0
[b] Sum of monounsaturated fatty acids: *cis*-C16:1 (*n*-7, *n*-9) + *cis*-C18:1 *n*-9 + *cis*-C18:1 *n*-7 + *trans*-C18:1 *n*-7 + C20:1
[c] Sum of polyunsaturated fatty acids: *cis*-C18:2 *n*-6 + *cis*-C18:3 *n*-3 + C20:3 *n*-6 + C20:4 *n*-6 + C20:5 *n*-3 + C22:4 *n*-6 + C22:5 *n*-3 + C22:6 *n*-3
[d] Ratio of sum polyunsaturated : sum of saturated fatty acids
[e] Sum of *n-3* fatty acids
[f] Sum of *n-6* fatty acids
[g] Ratio of sum of *n-3* : sum of *n-6*

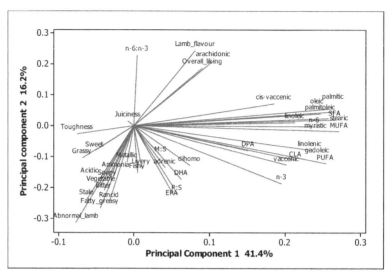

Figure 6.2 The dimensionally reduced fatty acid and taste panel data. The first component is a measure of total composition of fatty acids and the second component may interpreted as a measure of normal sensory variables (lamb flavour, overall liking) correlated with arachidonic acid (C20:4) and the ratio of *n-6: n-3*

6.3.3 Genetic Parameters

Univariate heritability estimates for eating quality traits are shown in Table 6.5. The heritabilities were quite variable, with acidic, bitter, fatty/greasy, stale, sweet and vegetable, being highly heritable. The heritabilities for juiciness, abnormal flavour and lamb flavour were moderate and ranged from 0.31 for juiciness to 0.21 for lamb flavour.

The estimated residual correlations among eating quality traits are presented in Table 6.6. The correlation for lamb flavour with almost all the adverse eating traits (abnormal flavour, acidic, ammonia and bitter flavour) was significantly negative, however it was positively correlated with juiciness. Also, juiciness was significantly positively correlated with fatty/greasy flavour, and negatively correlated with bitter flavour. Overall liking was significantly positively correlated with fatty/greasy, juiciness and lamb flavour (0.34, 0.25, and 0.62, respectively), and also was negatively correlated with abnormal flavour, ammonia, bitter and metallic flavour (-0.41, -0.27, -0.31, -0.20, respectively). Finally,

119

correlations of toughness with juiciness and lamb flavour were negative (-0.43 and -0.11). Thus, these high correlations suggest that fatty/greasy flavour, juiciness and toughness of meat have the greatest effect on lamb flavour.

Table 6.5 *Univariate heritabilities (h^2), with standard errors (s.e.) for sensory panel traits (0-100 scales)*

Trait	h^2	s.e.
Abnormal flavour	0.27	0.14
Acidic	0.68	0.14
Ammonia	0.25	0.13
Bitter	0.68	0.15
Fatty/greasy	0.57	0.19
Fishy	0.09	0.11
Grassy	0.12	0.11
Juiciness	0.31	0.17
Lamb flavour	0.21	0.13
Livery	0.22	0.13
Metallic	0.09	0.10
Overall liking	0.05	0.09
Rancid	0.30	0.16
Soapy	0.04	0.08
Stale	0.63	0.14
Sweet	0.92	0.14
Toughness	0.15	0.13
Vegetable	0.60	0.20

Table 6.6 Residual correlations for eating quality traits. Significant (P< 0.05) residual correlations (except diagonal) are shown in bold

Trait	Abnormal flavour	Acidic	Ammonia	Bitter	Fatty/greasy	Fishy	Grassy	Juiciness	Lamb flavour	Livery	Metallic	Overall liking	Rancid	Soapy	Stale	Sweet	Toughness	Vegetable
Abnormal flavour	1.00																	
Acidic	**0.36**	1.00																
Ammonia	**0.30**	0.08	1.00															
Bitter	**0.39**	-0.01	**0.20**	1.00														
Fatty/greasy	0.00	0.00	-0.19	0.07	1.00													
Fishy	**0.40**	**0.38**	0.09	0.02	**-0.20**	1.00												
Grassy	**-0.30**	-0.18	**-0.37**	**-0.47**	0.01	-0.02	1.00											
Juiciness	0.01	0.03	-0.08	**-0.25**	**0.22**	0.17	0.11	1.00										
Lamb flavour	**-0.58**	**-0.20**	**-0.29**	**-0.33**	0.05	-0.04	0.18	**0.22**	1.00									
Livery	**0.24**	**0.24**	**0.16**	**0.25**	0.17	-0.09	**-0.21**	0.03	-0.06	1.00								
Metallic	**0.23**	**0.29**	**0.11**	**0.21**	0.06	0.09	-0.19	0.10	0.02	0.09	1.00							
Overall liking	**-0.41**	-0.18	**-0.27**	**-0.31**	**0.34**	-0.02	0.09	**0.25**	**0.62**	-0.17	**-0.20**	1.00						
Rancid	0.18	0.19	0.17	**0.23**	0.03	-0.06	0.05	-0.01	**-0.22**	**0.47**	0.06	**-0.21**	1.00					
Soapy	**0.22**	**0.31**	0.15	0.12	0.18	**0.20**	-0.17	**0.29**	0.07	**0.25**	**0.40**	-0.02	0.08	1.00				
Stale	**0.21**	0.03	0.13	**0.47**	-0.01	0.04	**-0.25**	**-0.28**	-0.17	-0.06	**0.35**	**-0.29**	**0.24**	0.13	1.00			
Sweet	**-0.32**	-0.01	**-0.28**	**-0.66**	0.05	-0.09	0.18	**-0.43**	**0.29**	-0.19	**-0.29**	**0.38**	**-0.25**	0.04	**-0.34**	1.00		
Toughness	-0.07	-0.10	**-0.30**	0.05	0.01	0.06	-0.03	-0.09	-0.11	**-0.44**	0.07	-0.06	**-0.30**	-0.13	**0.23**	-0.08	1.00	
Vegetable	-0.16	-0.09	0.09	0.02	0.11	-0.08	-0.19		0.06	**-0.33**	**-0.32**	0.06	**-0.28**	-0.19	0.11	0.15	**0.34**	1.00

Residual correlations between eating quality traits and fatty acid composition (mg/100g muscle) are given in Table 6.7. The correlations were significantly negative for all the polyunsaturated fatty acids with lamb flavour, and arachidonic acid and EPA showed the highest correlation coefficients (-0.54 and -0.46, respectively). Correlations for mostly all fatty acids were negative with meat juiciness. However, correlations for fatty acids with toughness of meat were generally low and did not show a particular pattern. Finally, correlations for overall liking followed a similar pattern with lamb flavour. In particular, overall liking was negatively correlated with *cis*-vaccenic and all the individual polyunsaturated fatty acids.

Of interest were the positive and negative correlations of overall liking with the proportion of total fatty acids that were MUFA and PUFA respectively. These results imply that meat with high MUFA and low PUFA proportions will be perceived to taste better. The high correlations suggest that fatty acid composition affects flavour, juiciness and overall liking to an important extent.

Table 6.7 *Residual correlations between eating quality traits and fatty acid composition of intramuscular fat. Significant (p<0.05) residual correlations are shown in bold*

Traits	Lamb flavour	Juiciness	Toughness	Overall liking
Myristic acid - 14:0	-0.07	-0.13	0.02	-0.10
Palmitic acid - 16:0	0.02	-0.08	-0.12	0.07
Stearic acid - 18:0	-0.03	**-0.20**	-0.09	0.07
Palmitoleic acid – cis 16:1 (n-7, n-9)	-0.02	-0.11	-0.06	-0.06
Oleic acid – cis 18:1 n-9	0.00	-0.10	-0.13	0.10
Cis-Vaccenic acid – cis 18:1 n-7	-0.18	-0.19	0.05	**-0.39**
Vaccenic acid – trans 18:1 n-7	0.10	-0.19	-0.04	0.05
Gadoleic acid - 20:1	0.04	-0.04	-0.13	-0.01
Linoleic acid – cis 18:2 n-6	**-0.25**	-0.12	-0.05	**-0.31**
Linolenic acid – cis 18:3 n-3	-0.19	**-0.37**	0.01	**-0.23**
Dihomo-γ-linolenic acid - 20:3 n-6	**-0.42**	0.01	0.04	**-0.43**
Arachidonic acid - 20:4 n-6	**-0.54**	-0.18	0.12	**-0.60**
EPA (Eicosapentanoic acid) - 20:5 n-3	**-0.46**	-0.11	0.15	**-0.50**
Adrenic acid - 22:4 n-6	**-0.21**	**-0.27**	0.02	**-0.34**
DPA (Docosapentaenoic acid) - 22:5 n-3	**-0.33**	**-0.32**	0.08	**-0.50**
DHA (Docosahexaenoic acid) - 22:6 n-3	**-0.29**	-0.09	0.25	**-0.24**
Conjugated linoleic acid (CLA)	0.03	-0.13	-0.03	-0.01
% SFA	**0.25**	**0.40**	-0.07	**0.25**
% MUFA	**0.30**	**0.25**	-0.11	**0.29**
% PUFA	**-0.45**	**0.31**	0.11	**-0.54**

6.4 Discussion

The dataset analysed provided us with a unique opportunity to investigate, in detail, the relationship between taste panel traits and fatty acid composition. Selection for increased leanness did not change many aspects of eating quality, although meat from the FAT line was perceived to be juicier than meat from LEAN line. The heritability estimates for most of the eating quality traits were variable, with the exception of fishy, grassy, metallic, overall liking, soapy, and toughness, which had estimates that were small and not significantly different from zero.

Due to a lack of information on heritabilities in eating quality traits in sheep, we will compare our results with published studies in other species. The estimate of heritability for toughness (0.15) from the current analyses is similar to the heritability of 0.23, which is reported by Wilson *et al.* (1976) for polled Hereford sires and Angus-Holstein cows. Most previous research (*e.g.* Van Fleck *et al.*, 1992; Barkhouse *et al.*, 1996; Splan *et al.*, 1998; Nephawe *et al.*, 2004) indicated that selection for decreased toughness would result in little, if any, genetic progress. The reported estimate for this trait was 0.10 from crossbred cattle (Van Fleck *et al.*, 1992), 0.06 from crossbred steers and heifers (Barkhouse *et al.*, 1996), 0.05 from crossbred steers and heifers (Spaln et al, 1998), and 0.26 from steers (Nephawe *et al.*, 2004).

Our heritability estimate for juiciness was moderate (0.31), and in close agreement with Wilson *et al.* (1976), that estimated a heritability of 0.26. In contrast, other studies in beef cattle reported estimates of 0.14, 0.0, and 0.05 (Van Fleck *et al.*, 1992; Splan *et al.*, 1998; Nephawe *et al.*, 2004, respectively).

Flavour, is assumed to have low heritability, based on estimates from the literature. Lamb flavour, in our study, was moderately heritable (0.21), which is in contradiction with other estimates in studies in beef cattle. Specifically, the heritabilities of beef flavour were 0.03 (Van Fleck *et al.*, 1992), 0.04 (Splan *et al.*, 1998), and 0.05 (Nephawe *et al.*, 2004).

In pigs, Cameron (1990) reported heritability estimates for toughness, pork flavour, juiciness and overall liking of 0.23, 0.16, 0.18 and 0.16, respectively, from data on 40 full-sib litter groups of Duroc and halothane-negative British Landrace pigs, in agreement with estimates of 0.15 for toughness and 0.21 for lamb flavour, in the present study. In contrast,

Lo *et al.* (1992) reported heritability estimates of 0.45, 0.13, 0.12 and 0.34 for toughness, pork flavour, juiciness and overall liking, respectively, from data on Duroc and Landrace pigs. Based on results of a small number of studies, the same eating quality traits in general seemed to be low to moderately heritable of the order of 0.10 to 0.20 (Lo, 1990), in agreement with the present study. However, these estimates tend to be quite variable, as would be expected given the relatively small sample sizes generally used in these studies. Overall, these results indicate that sensory meat quality traits, assessed by taste panels, are determined to some extent by additive genetic effects and as such there is some scope for genetic improvement by means of selection.

6.4.1 Relationships between sensory scores and fatty acid concentrations

The residual correlations were used as an approximation of phenotypic correlations. Note that a sire model was used so it would have been possible to present phenotypic correlations by adding sire components to environmental components. The correlations were strong for most of the polyunsaturated fatty acids with lamb flavour, overall liking and juiciness. The high correlations suggest that fatty acid composition affects flavour development and juiciness of lamb meat to an important extent.

Correlations between lamb flavour and polyunsaturated fatty acids were strongly negative. Other reports in the literature (Kemp *et al.*, 1981; Larick and Turner, 1990; Melton, 1990) have shown that the *n*-6 and *n*-3 polyunsaturated fatty acids are important contributors to the flavour of ruminant meats fed grain or grass-fed. The development of rancid flavours in grass-fed lambs comes from the degradation of polyunsaturated fatty acids in the tissue membranes. This might explain the negative correlations of flavour and overall liking with polyunsaturated fatty acids.

The strong positive correlation between overall liking and proportion of total MUFA is of particular interest, considering the benefits of MUFA in human diet. The study by Etherton *et al.* (1999) reported that high-MUFA diets lowered total cholesterol in humans by 10% and LDL ('bad') cholesterol by 14%. Also, in the same study triacylglycerol concentrations were 13% lower in subjects consuming the high-MUFA diets. Although the cholesterol-lowering response to PUFA is greater than to MUFA, there has been caution in recommending high PUFA diets, because of potentially adverse health effects of their lipoperoxidation products (Williams, 2000). Recent studies show that low fat diets can raise triglycerides and reduce the levels of protective HDL cholesterol (Mensink *et al.*,

1987; Sandstrom *et al.*, 1992), have renewed interest in the possibility of altering fat quality, by altering fatty acid composition of animal products, as a cholesterol-lowering strategy within the population.

6.5 Conclusions

In summary this study has shown that altering carcass fatness has simultaneously changed some aspects of eating attributes, with meat from FAT line animals being juicier and, less easily explicable, having a significantly more vegetable flavour. Heritabilities for most of the eating quality traits were moderate to high, suggesting that lamb meat perception can be changed genetically, jointly with fatty acid composition of meat. It has verified that some taste panel traits are strongly associated with fatty acid profile.

Chapter Seven
General Discussion

7.1 Objectives overview

Changes to the national and international meat market have brought new challenges for lamb meat production. This is not only because lamb consumption has declined, but also because the consumers are becoming increasingly conscious of product quality. High expectations of meat quality and for healthy meat products by consumers are forwarded to the meat processing industry and thus to the sheep breeders. The expectation of maintained high meat quality increases the interest for investigating the factors that influence meat quality. Meat does not have a predetermined quality; many factors can affect meat quality – from genetics, to growth and slaughter and the final product. Numerous features, both genetic and environmental, are involved in the final meat quality. This thesis has focused on investigating the genetic background of meat quality.

As already mentioned throughout the thesis, the term meat quality concerns both the meat as a product (product quality) and the way the meat is produced (production quality) (Hofmann, 1994). The meat processing industry might be most concerned with technological meat quality, whereas the consumer is concerned with sensory meat quality and, increasingly, production quality (Hofmann, 1994).

Prior to this study, the benefits of new measurement technology, such as CT, that offers more accurate measurements of carcass traits over other *in vivo* technology, had not been thoroughly quantified in sheep. Furthermore, the question of prediction of the genetic relationship between carcass and meat quality traits by CT was unknown.

With regard to meat quality, little was known on both quantitative genetics and identification of QTL for meat quality traits in sheep. This was defined as a research objective for this thesis and was perceived as a relevant requisite for allowing the development of an appropriate framework for predicting meat quality in the live animal. The relevance of this study was that, for the first time, this thesis presented predictive relationships between CT measurements and meat quality attributes. Therefore, this thesis

126

has focused on the understanding of the quantitative genetics of both CT assessed and meat quality traits, and also in identifying QTL for meat quality in sheep.

7.2 Key findings

The main developments and findings of each of the five research chapters are summarised below.

Chapter 2: This Chapter verified that CT can be used to genetically improve carcass composition and conformation, under field conditions. After five years of selection on an index designed to improve both composition and conformation (the 'CT index'), a large response was observed in the CT index, with genetic progress of 52 units/year, equivalent to 0.11 phenotypic standard deviations per year. Heritabilities for the index, and the component traits of average CT-assessed muscle area, ultrasonic muscle depth and ultrasonic fat depth were 0.41 (s.e. 0.08), 0.38 (s.e. 0.07), 0.41 (s.e. 0.05) and 0.30 (s.e. 0.05), respectively. The index was positively genetically correlated with ultrasonic muscle depth and carcass weight and negatively genetically correlated with fat class. The genetic and phenotypic correlations among ultrasonic measurements were positive and moderate. However, many of the genetic correlations tended to have large standard errors. Selection on CT index moderately improved conformation and was successful at decreasing fat class of the carcass. Thus, genetic improvement of carcass quality can be achieved in hill sheep using CT assessed traits.

Chapter 3: Genetic parameters were presented for carcass composition and meat quality traits in Scottish Blackface sheep. CT was used to obtain non-destructive *in vivo* estimates of the carcass composition of 700 lambs, at *ca.* 24 weeks of age, with tissue areas and image densities obtained for fat, muscle and bone components of the carcass. Meat quality was assessed on 350 male lambs, at *ca.* 8 months of age, which had previously been CT scanned. Meat quality traits included intramuscular fat content, initial and final pH of the meat, colour attributes, shear force, dry matter, moisture and nitrogen proportions, and taste panel assessments of the cooked meat. FAT line animals were significantly (p<0.05) fatter than the LEAN line animals in all measures of fatness (from CT and slaughter data), although the differences were modest and generally proportionately less than 0.1. Correspondingly, the LEAN line animals were superior to the FAT line animals in muscling measurements. Compared to the LEAN line, the FAT line had lower muscle density (as indicated by the relative darkness of the scan image), greater estimated subcutaneous fat (predicted from fat classification score) at slaughter, more intramuscular fat content, a more 'yellow' as opposed to 'red' muscle colour, and juicier meat (all $p<0.05$). All CT tissue areas were moderately to highly heritable, with h^2 values ranging from 0.23 to 0.76. Likewise, meat quality traits were also moderately heritable. Muscle density was the CT trait most consistently related to meat quality traits, and genetic correlations of muscle density with live weight, fat class, subcutaneous fat score, dry matter proportion, juiciness, flavour and overall liking were all moderately to strongly negative, and significantly different from zero. In addition, intramuscular fat content was positively genetically correlated with juiciness and flavour, and negatively genetically correlated with shear force value. The results of this study demonstrate that altering carcass fatness will simultaneously change muscle density (indicative of changes in intramuscular fatness), and aspects of intramuscular fat content, muscle colour and juiciness. The heritabilities for the meat quality traits indicate ample opportunities for altering most meat quality traits. Moreover, it appears that colour, intramuscular fat content, juiciness, overall liking and flavour may be adequately predicted, both genetically and phenotypically, from measures of muscle density. Thus, genetic improvement of carcass composition and meat quality is feasible using *in vivo* measurements.

Chapter 4: Quantitative trait loci (QTL) were identified for traits related to carcass (600 lambs) and meat quality (300 lambs) in nine half-sibs from the same population described in **Chapter 3**. In total, nine genome-wise significant and 11 chromosome-wise and suggestive QTL were detected in seven out of eight chromosomes. Genome-wise significant QTL were mapped for lamb flavour (OAR 1); for muscle densities (OAR 2 and OAR 3); for colour a* (redness) (OAR 3); for bone density (OAR 1); for slaughter live weight (OAR 1 and OAR 2) and for the weights of cold and hot carcass (OAR 5). The QTL with the strongest statistical evidence affected the lamb flavour of meat and was on OAR 1, in a region homologous with a porcine SSC 13 QTL identified for pork flavour. This QTL segregated in 4 of the 9 families. This study provided new information on QTL affecting meat quality and carcass composition traits in sheep.

Chapter 5: Genetic parameters for *longissimus* muscle fatty acid composition were estimated in the population (350 lambs), described in **Chapter 3**. Furthermore, quantitative trait loci (QTL) were identified for the same fatty acids using data on 300 lambs. Fatty acid composition measurements included in total 17 fatty acids of three categories: saturated, monounsaturated, and polyunsaturated. The FAT line had a greater intramuscular fat content and more oleic acid, but less linoleic acid (18:2 *n*-6) and DPA acid (22:5 *n*-3) than did the LEAN line. Saturated fatty acids were moderately heritable, ranging from 0.19 to 0.29 and total saturated fatty acids (SFA) were highly heritable (0.90). Monounsaturated fatty acids were moderately to highly heritable, with *cis*-vaccenic acid (18:1 *n*-7) being the most heritable (0.67), and total monounsaturated fatty acids (MUFA) were highly heritable (0.73). Polyunsaturated fatty acids were also moderately to highly heritable with arachidonic acid (20:4 *n*-6) and conjugated linoleic acid (CLA) being the most heritable, with values of 0.60 and 0.48, respectively. The total polyunsaturated fatty acids (PUFA) were moderately heritable (0.40). In total, 21 chromosome-wide QTL were detected in six out of eight chromosomal regions. The chromosome-wide significant QTL affected three saturated, five monounsaturated, and 13 polyunsaturated fatty acids. The most significant result was a QTL affecting linolenic acid (18:3 *n*-3) on chromosome 2. This QTL segregated in 2 of the 9 families and explained 37.6% of the phenotypic variance. Also, 10 significant QTL were identified on chromosome 21, where 8 out of 10 QTL were segregating in the same families and detected at the same position. The results of this study demonstrate that altering carcass fatness will simultaneously change intramuscular fat content and oleic, linoleic, and DPA acid content. The heritabilities of the fatty acids indicate opportunities for genetically altering most fatty acids. Moreover, this is the first study of detection of QTL directly affecting fatty acid composition in sheep.

Chapter 6: Further analyses were performed for the eating quality traits measured in this population. Also, relationships between eating quality assessments and fatty acid composition were investigated. Eating quality measurements included 18 definitions. The FAT line had juicier meat and more vegetable flavour than the LEAN line. Heritabilities for most of the eating quality traits were variable, ranging from 0.21 (lamb flavour) to 0.92 (sweet flavour). Lamb flavour, juiciness and overall liking were strongly negatively correlated with individual polyunsaturated fatty acids, with the correlations being significantly different from zero. Overall liking was strongly positively correlated with the proportion of total monounsaturated fatty acids.

To summarise, this thesis has contributed substantial information to the genetics of sheep meat quality. It has been shown:
- ✓ There is substantial genetic variation in meat quality attributes, particularly in fatty acid composition of meat
- ✓ Meat quality attributes could be predicted *in vivo* by CT-muscle density
- ✓ Many QTL have been identified for a range of meat and carcass traits
- ✓ There is a strong relationship of fatty acid composition with flavour perceptions and juiciness of meat.

7.3 Implications – *Selection for meat quality*

Product quality data is expensive and difficult to collect. With varying definitions of quality, the benefit of building a selection and production programme around consumer product quality is difficult to identify. All traits that have been the focus of genetic improvement programmes so far have been traits that can be measured easily and relatively cheaply on the live animal either directly or indirectly using non-invasive technology, such as ultrasound and CT. Hence, this system works well with our traditional traits but it is time to look to the future and determine how best to work on traits that are much more closely related to meat quality.

Problems associated with improvement of carcass and meat quality traits include:
1. Carcass and meat quality traits of primary importance to sheep retailers and consumers are generally not included in sheep carcass grading and classification, therefore do not, currently, affect the income of commercial sheep farmers.
2. Most meat quality traits cannot be evaluated accurately in breeding animals.
3. It becomes very difficult to motivate breeders to invest in selection programmes to improve carcass and meat quality traits. However benefits from selection accrue many years in the future, therefore any selection programme initiated should be aimed at improving future industry profitability.
4. Due to lack of motivation, because of the long term nature of selection programmes, breeders may often lose interest before the real benefits arise.
5. Mode of improvement: should traits be best selected using live animal measures of CT or ultrasound, progeny testing, measures on indicator traits or gene mapping? What technologies are currently available and which are the most cost effective and accurate?

Thus, this section will focus on the potential opportunity to include meat and eating quality traits in a breeding programme using CT measured traits and QTL identified throughout this thesis, so that both quantity of product and quality of product can be increased. The general question to be addressed is *"How to select for meat and eating quality?"*

7.3.1 Breeding goals

A fundamental question that needs to be addressed is *"Which genetic traits can be selected for (or altered) at the genome level to satisfy the consumer's sensory and/or organoleptic and healthiness requirements without impairing efficiency in the livestock production chain?".*

Meat quality today, is not only about improving organoleptic traits (juiciness, flavour and intramuscular fat) but also about healthiness of meat. Consumer surveys world-wide have demonstrated that juiciness, flavour and intramuscular fat are the most important sensory quality attributes of meat, irrespective of animal species (Schönfeld, 2001). From the perspective of healthiness of meat, saturated fatty acid content of the human diet continues to be of concern for health-conscious consumers, especially with regard to plasma cholesterol concentration. The reason for concern is that cardiovascular disease is the leading cause of death (American Heart Association, 1999). For example, nearly 60 million people in the United States (20% of the population) have at least one type of cardiovascular disease such as hypertension, coronary artery disease, stroke, or rheumatic heart disease (American Heart Association, 1999). Animal products provide collectively 56% of total fat, 74% of saturated fatty acids, and 100% of the cholesterol consumed by humans (Rhee, 1992). Replacing saturated fatty acids with unsaturated fatty acids (MUFA and PUFA) rather than with carbohydrates is the common approach to normalize plasma lipid profiles (National Cholesterol Education Program, 2001). Diets rich in monounsaturated fatty acids (*e.g.*, Mediterranean diet) result in lipoprotein profiles that are more favourable for maintaining vascular health (Rivellese *et al.*, 2003, Kok and Kromhout, 2004). In summary, decreasing the saturated fatty acid composition of lamb meat by replacing these "less healthy" fatty acids with MUFA and PUFA would lead to an improvement in the healthiness of lamb meat products for use in the human diet.

The rationale behind this study is the philosophy that breeding goals of the future must reconcile meat quality, genetics and the consumer. This thesis showed that lamb meat quality is affected at the genetic level; hence it must be addressed as an integrated approach. A further question that needs to be answered is the following: *which one (or how many) of the following five dimensions, should the breeding goal actually address?*

i)   structuring breeding goals to satisfy the present consumer or the consumer of the future

ii)   structuring breeding goals to satisfy the producer and/or the commercial producer

iii) structuring breeding goals to satisfy the slaughterhouses and processors

iv)   structuring breeding goals to satisfy all the links in the supply chain

According to Grunert *et al.* (1997), the information on the end user's needs and trends is crucial. The value of product (as perceived by the end user) sets the limit for the price of a product and therefore the returns (earnings) for the entire value chain. Van Trijp, Steenkamp and Candel (1997) indicated a positive relationship between perceived quality and economic returns. The higher the perceived quality of a product, the higher is the selling price resulting in an increased market share and profitability.

Consequently, inclusion of traits in the breeding goal should be viewed within the context of the breeding programme and the broader sheep industry. This **Chapter** explores selection methods, using *in vivo* predictors and *"known"* QTL affecting meat quality, using selection index techniques. The following aspects were included as the breeding goals:

i)   **Health** of meat has evolved over the last decade especially as a major issue for the consumer. Thus, health *per se* must feature as a building block in the breeding goal. From a genetic point of view, identification of genes that influence *e.g.* fatty acid composition of meat could enhance the improvement of health.

ii)   **Eating quality** of meat, as preferred by consumers, such as flavour of meat. Any objectionable flavour can influence consumption and affect total consumption of meat.

iii) **Shelf-life** of meat, the development of oxidative off-flavours (rancidity) has been recognised as a serious problem during holding or storage of meat and thus the meat becomes unacceptable by the consumer. Altering the PUFA content in sheep may have important implications for meat quality, such as shelf-life characteristics of meat because of their greater susceptibility to oxidative breakdown and the production of volatile compounds during cooking. However, attempts to increase levels of PUFA in meat for health reasons could be compromised by increased lipid oxidation.

Table 7.1 *Breeding goals which are recommended for the meat sheep industry*

Trait	Reason for inclusion
Intramuscular Fat (IMF)	• Could be predicted by CT-muscle density (r_g= -0.67) • Affects juiciness and flavour of meat positively (r_g= 0.69, 0.52, respectively) • Heritability of this trait is moderate (0.32)
Flavour	• Most important sensory trait for the consumer • One of the primary consumer acceptance criteria of lamb
Juiciness	• It affects the consumer's impression and acceptance of lamb • Positive relationship with IMF and flavour
Fatty acid profile	• PUFA: beneficial effects on human health, but susceptible to oxidation with an effect on meat shelf-life • MUFA of the *cis*-configuration protect against coronary heart disease • Could be predicted by CT-muscle density

7.3.2 Defining selection index calculations

7.3.2.1 Data and genetic parameters

In order to calculate the properties of the selection indices, genetic and phenotypic parameter estimates (heritabilities and correlations) are required for all goal and index traits. The genetic and phenotypic parameter values were taken from the results presented in this thesis. In the cases where heritabilities were very high (*e.g.* > 0.50) a value of 0.50 was used instead. Univariate heritabilities and phenotypic standard deviations used in index calculations are shown in Table 7.2.

Table 7.2 *Heritabilities (h^2), phenotypic standard deviations (σ_p) and mean values used in the index calculations*

Trait	h^2	σ_p	mean
Intramuscular fat (IMF) (mg/100g)	0.32	815	2511
Flavour (0-100 units)	0.11	4.88	26.5
Juiciness	0.21	5.92	41.3
Live weight at CT (LW) (kg)	0.20	3.94	32.2
Muscle density at CT (MD)	0.35	2.21	43.3
SFA (mg/100g)	0.50	361	987
MUFA (mg/100g)	0.50	301	984
PUFA (mg/100g)	0.40	42.4	233
Conjugated linolenic acid (CLA) (mg/100g)	0.48	15.2	33.4
Linolenic acid (mg/100g)	0.30	12.6	39.2
Oleic acid (mg/100g)	0.27	286	828

The genetic and phenotypic correlations used in the index calculations are shown in Table 7.3. The same parameters were used for each selection index.

Table 7.3 *Correlations[†] between breeding goal and index traits*

	IMF	Flavour	Juiciness	LW	MD	SFA	MUFA	PUFA	CLA	Linolenic acid	Oleic acid
IMF		0.20	0.12	0.23	-0.57	0.71	0.40	0.33	0.40	0.40	0.40
Flavour	0.52		0.22	-0.45	-0.20	-0.20	0.30	-0.45	0.00	-0.19	0.00
Juiciness	0.69	0.22		0.00	-0.16	-0.13	0.25	-0.31	-0.13	-0.37	-0.10
LW	0.38	-0.17	0.00		0.20	0.14	0.29	0.26	-0.05	0.00	0.00
MD	-0.67	-0.50	-0.50	0.20		-0.30	-0.35	-0.26	-0.15	0.00	0.00
SFA	0.71	-0.20	-0.13	0.14	-0.45		0.49	0.28	0.87	0.55	0.97
MUFA	0.40	0.30	0.25	0.23	-0.50	0.49		0.26	0.50	0.52	0.90
PUFA	0.33	-0.45	-0.31	0.50	-0.56	0.28	0.26		0.16	0.50	0.26
CLA	0.40	0.00	-0.13	0.00	-0.50	0.87	0.16	0.50		0.37	0.50
Linolenic acid	0.40	-0.19	-0.37	0.00	-0.24	0.55	0.50	0.52	0.37		0.52
Oleic acid	0.40	0.00	-0.10	0.00	-0.30	0.97	0.26	0.50	0.50	0.50	

[†]Phenotypic correlations above and genetic correlations below the diagonal.

7.3.2.2 Relative economic values

In all cases described below, the selection goals were restricted to single traits or pairs of traits. Therefore, relative economic values of +/- 1 were used, simply to indicate whether the goal traits should be increased or decreased. A full economic appraisal of the traits is beyond the scope of this thesis.

7.3.2.3 Index equations

A selection index combines information from an individual's own and relatives' phenotypic performance for multiple traits into an overall score. It is an effective way of selecting breeding stock when several traits are being evaluated. The best way to implement multiple trait selection is *via* EBVs from best linear unbiased prediction (BLUP). However, a multi-trait BLUP framework does not readily lend itself to decision making regarding which traits and measurements to include in an index and the relative importance of each trait (Conington *et al.*, 2001). This may be achieved by selection index theory. In this case, selection indices that mimic the typical contribution from relatives with BLUP may be constructed to: i) determine how each trait contributes to selection, ii) determine which traits are important and which can be dropped from the index with little effect, iii) calculate the expected genetic change in each trait with selection, and iv) determine the overall accuracy of selection.

The selection index used the following assumed data structure for the traits: i) a measurement on the lamb itself, ii) a measurement on a *"known"* QTL for the goal trait, assuming 5, 10, 20 or 50% genetic variance explained by the QTL, and iii) a measurement on 30 paternal half-sibs of the lamb.

The selection index calculations were programmed using *GENSTAT* (2003) and based on index theory outlined by Cunningham (1969). The index matrices and vectors were:

X = vector of phenotypic observations for the m variables (measured traits) or sources of information included in the index; v = vector of relative economic weights of the n traits included in the breeding goal; b = vector of weighting factors to be used in the index; P = m x m matrix of phenotypic covariances between the m measured variables in X. G = m x n matrix of genetic covariances between the m measured variables in X and the n traits in the breeding goal. $C = n \times n$ matrix of genetic covariances between the n traits in the breeding goal; D = selection differential on a standardized normal distribution.

The general solution to the index equations is:

$$b = P^{-1}Gv$$

The variance of the index is:

$$b'Pb$$

and the accuracy (r_I) of the index is:

$$\sqrt{\frac{b'Pb}{v'Cv}}$$

To derive the progress in each trait, a selection intensity of 1 was assumed. The genetic gain for each trait per generation is the regression of each trait in the aggregate genotype on the index, *i.e.* the ith element of:

$$\frac{b'G}{b'Pb}$$

One round of selection on the index will produce $D_{\sigma I}$ units of change in the index, and therefore a vector:

$$D \frac{b'G}{\sqrt{b'Pb}}$$

of units of change in each trait.

The contribution of each index trait to the breeding goal was calculated as the proportional reduction of the response in the breeding goal, if the trait was excluded from the index. Therefore, the contribution of a trait in the index to the breeding goal is actually measured as the proportional reduction to the accuracy of the index if the trait was excluded from the index. The contribution of a trait or measurement in the index to the breeding goal is:

$$\frac{r_I^*}{r_I} = 1 - \sqrt{1 - \frac{b_j^2}{b' PbP_{jj}^{-1}}}$$

Where, r_I^* is the accuracy of the selection index with the trait X_I omitted from the selection index and P_{jj}^{-1} is the jth diagonal element of the inverse of P the matrix (Cunningham, 1975).

7.3.2.4 Breeding schemes with *"known"* QTL

In the breeding schemes with QTL assisted selection, additional information, consisting of information on a *"known"* QTL, was available for an index based on phenotypic information of the candidate. The QTL explained 5, 10, 20 or 50% of the genetic variance of the trait of interest, referred to as Q05, Q10, Q20 and Q50, respectively. The remaining genetic variation resulted from polygenes (*i.e.* not marked). In the index calculations, the QTL information was modelled as a trait that was correlated with the breeding goal trait and had a heritability of 1. The correlations between the QTL and the breeding goal trait depended on the amount of variation that was explained by the QTL. Genetic correlations of the breeding goal trait with the QTL were $\sqrt{q}$, where q is the fraction of the genetic variance explained by the QTL.

7.3.3 Index calculation results

The results from this work (presented below) show that selection indices can theoretically improve meat flavour, meat profile into a more healthy option for human, and meat shelf-life for the benefit of the market.

Genetic improvement of sheep meat has largely focused on increasing growth rate and muscularity while reducing total carcass fat content in sheep. This trend to reduce the total amount of carcass fat is mainly due to the consumer demand and price incentive (for producers) for lean carcass. However, we have shown that flavour and juiciness of sheep meat is related to its intramuscular fat (IMF) content and the industry aim is to retain IMF at 3-5g/100g meat (S. Kitessa, personal communication). Intramuscular fat in sheep is also an important source of essential fatty acids that have health benefits. This section presents whether breeding goals offer opportunities to: i) increase the perception of meat flavour and juiciness and retain the IMF content from the consumer perspective, ii) increase the proportion of the beneficial fatty acids while reducing the amount of saturated fatty acids in sheep meat from the healthiness perspective, and iii) increase the proportion of beneficial fatty acids and protect those lipids from pro-oxidative environment from the market perspective.

7.3.3.1 Expected benefits from selection for meat flavour

Three breeding goal traits are evaluated for their inclusion into the selection indices to improve meat flavour. They include intramuscular fat, flavour and juiciness of meat. To achieve the goals, five recorded ('index') traits are considered. The index traits are live weight and CT-muscle density measured on the individual animal and combinations of measures on 30 paternal half-sibs or *"known"* QTL for the goal trait (*e.g.* flavour) each time explaining 5, 10, 20 or 50% of the genetic variance of the goal trait. The goal and index traits are shown in Table 7.4.

Table 7.4 *Breeding goals and index (measured) traits to improve meat flavour*

Breeding goal traits	Index traits
Intramuscular fat (IMF)	Live weight on individual and half-sibs + CT-muscle density on individual and half-sibs
IMF + Flavour	Live weight on individual and half-sibs + CT-muscle density on individual and half-sibs + *"known"* QTL for flavour explaining 5, 10, 20 or 50% of the genetic variance of the goal trait (flavour)
IMF + Juiciness	Live weight on individual and half-sibs + CT-muscle density on individual and half-sibs
Flavour	Live weight on individual and half-sibs + CT-muscle density on individual and half-sibs + *'known"* QTL for flavour explaining 5, 10, 20 or 50% of the goal trait (flavour)

The accuracy of each index was higher when combinations of measures on half-sibs or QTL were used, and the results are shown in Table 7.14 (**Appendix II**). Selection on CT-muscle density (highly correlated with IMF content) had a predicted progress in IMF content, ranging from 8.65 – 9.81% of the mean, using only individual measurement or measurement on 30 paternal half-sibs, respectively (Table 7.5). In the case where the selection goal was to improve IMF content and flavour, when a *"known"* QTL for flavour was included in the index, the expected progress in IMF was higher when the QTL for flavour explained 50% of the genetic variance of the trait, (13.7% mg/100g). However, in a real situation, a QTL is unlikely to explain 50% of the genetic variance, and hence using a QTL which explains less than 20% of the genetic variance gave the same response as using information on relatives. In the same index, the small responses in flavour, in the desired direction, reflect the low heritability for this trait. In the third index case explored, the goal was to improve IMF and at the same time increase juiciness. The responses were in the same range for IMF, as in the previous situations and juiciness increased by 4.35% units proportional of the mean, when 30 paternal half-sibs were used. In the final case, where the selection goal was to improve flavour perception, the progress ranged from 1.82% - 4.36% of the mean, for individual and QTL information.

Table 7.5 *Expected genetic change in each goal trait to improve meat flavour[‡]*

Response in	Economic value	IND[†]	IND+HS30	IND+Q05	IND+Q10	IND+Q20	IND+Q5
IMF (mg/100g) (LW+MD)*	+1	8.65	9.81				
IMF (mg/100g) + **Flavour** (units 0-100) (LW+MD+QTL_FLAVOUR)*	+1 +1	8.64 1.50	9.82 1.51	9.14 1.89	9.68 1.90	10.7 2.65	13.7 4.16
IMF(mg/100g) + **Juiciness** (units 0-100) (LW+MD)*	+1 +1	8.64 4.12	9.82 4.35				

[†]The genetic change expressed as % of the mean of the goal trait.
[‡]To derive the genetic progress in each trait for one round of selection, a selection intensity of 1 was assumed.
*Measured ('index') traits to achieve each goal.

The importance of each measured trait in the overall index is shown in Table 7.6. The figures shown represent the proportional genetic improvement lost if a particular measurement was dropped from the index, compared with the genetic gain achieved by the full index. Therefore, the higher the value, the more important that measurement is and the more it contributes to the index. For example, for Index 1 where we assume information only on the individual, if CT-muscle density is dropped from the index, the genetic progress would be reduced by 64%. The results show that CT-muscle density and a *"known"* QTL, which explains more than 20% of the genetic variance, are the most important traits for all the indices. This is due to the relatively high correlations between intramuscular fat, flavour, and juiciness with live weight and CT-muscle density traits, as well as the high heritability estimate for IMF (0.32). Live weight is the least important trait for the indices.

Table 7.6 *Contribution of each trait to the index*[†]

		IND	*IND+HS30*	*IND+Q05*	*IND+Q10*	*IND+Q20*	*IND+Q50*
Index 1	**LW +**	0.16	0.05				
	MD (IMF)*	0.64	0.08				
	LW-HS		0.08				
	MD-HS		0.25				
Index 2	**LW +**	0.15	0.05	0.13	0.12	0.08	0.03
	MD +	0.64	0.08	0.44	0.32	0.18	0.02
	QTL$_{FLAVOUR}$			0.06	0.11	0.19	0.37
	(IMF+Flavour)*						
	LW-HS		0.09				
	MD-HS		0.24				
Index 3	**LW +**	0.14	0.08				
	MD +	0.65	0.09				
	(IMF+Juiciness)*						
	LW-HS		0.05				
	MD-HS		0.24				
Index 4	**LW +**	0.02	0.01	0.02	0.03	0.04	0.06
	MD +	0.74	0.08	0.02	0.07	0.19	0.33
	QTL$_{FLAVOUR}$			0.15	0.25	0.38	0.58
	LW-HS		0.02				
	MD-HS		0.29				

[†] The higher the value, the more the trait contributes to the index.
*Measured ('index') traits to achieve each goal.

7.3.3.2 Expected benefits from selection for meat healthiness

Like any other food, meat and meat products contain elements which in certain circumstances, and in inappropriate proportions, have a negative effect on human health. Fatty acid composition has a considerable effect on the diet/health relationship, since each fatty acid affects the plasmatic lipid differently. Meat lipids usually contain less than 50% SFA (SFA of which only 25-35% have atherogenic properties), and up to 70% (lamb 50-52%, beef 50-52%, pork 55-57%, chicken 70%) unsaturated fatty acids (MUFA and PUFA) (Romans *et al.*, 1994). The presence of MUFA and PUFA in the diet reduces the level of plasma low-density lipoproteins-cholesterol (LDL), although PUFA also depress the high density lipoproteins-cholesterol (HDL) (Mattson and Grundy, 1985). Hence, it does not seem reasonable to consider meat as highly saturated food, especially in comparison with some other products (*e.g.* some dairy products).

Current methods of increasing the content of the beneficial fatty acids (PUFA) of meat have mainly focused on feeding sheep food supplements and additives. In addition, a generation of cloned transgenic pigs rich in omega-3 fatty acids has been created (Lai *et al.*, 2006). Hence, this section explores the possibility of appropriate selection goals that can change the total fatty acid composition in sheep for the benefit of the consumer, without the need for supplements or any kind of transgenic modification. The focus was on: i) increasing total MUFA of meat and at the same time not altering meat flavour, ii) increasing beneficial fatty acids (*e.g.* *cis*-CLA, linolenic acid), and iii) increasing, particularly, oleic acid which has beneficial effects on human health (**Chapter 5**). Specifically, the five goal traits are used and the index measures are live weight and CT-muscle density measured on the individual animal and combinations of measures on 30 paternal half-sibs or *"known"* QTL for the goal trait (*e.g.* CLA, oleic, linolenic) each time explaining 5, 10, 20 or 50% of the genetic variance of the goal trait. The goal and index traits are shown in Table 7.7.

Table 7.7 *Breeding goals and index (measured) traits to improve meat healthiness*

Breeding goal traits	Index traits
Flavour + MUFA	Live weight on individual and half-sibs + CT-muscle density on individual and half-sibs
CLA	Live weight on individual and half-sibs + CT-muscle density on individual and half-sibs + *"known"* QTL for CLA explaining 5, 10, 20 or 50% of the genetic variance of the goal trait (CLA)
Oleic acid	Live weight on individual and half-sibs + CT-muscle density on individual and half-sibs + *"known"* QTL for oleic acid explaining 5, 10, 20 or 50% of the genetic variance of the goal trait (oleic)
Flavour + Linolenic acid	Live weight on individual and half-sibs + CT-muscle density on individual and half-sibs + *"known"* QTL for linolenic acid explaining 5, 10, 20 or 50% of the genetic variance of the goal trait (linolenic)

The fatty acid composition of lamb meat can be altered through traditional selection to meet consumer demands for healthful fats (*i.e* lower SFA and higher MUFA or PUFA). In general, greater responses are seen for total MUFA, CLA, oleic and linolenic acid (Table 7.8). All predicted responses are in the desired direction, with small improvement in flavour perception. In practice, flavour will be held more or less constant whilst improvements in MUFA, CLA, oleic and linolenic acid are achieved. In addition, the genetic change gained from information using *"known"* QTL, the four index cases that investigated, gave a higher response even when QTL explained 5% of genetic variance of the goal trait. Furthermore, the change in SFA when index 1 is used, although SFA are not included in the selection goal, can be determined by the correlated response in SFA, which is 7% of the mean. In the situation that we use a selection index for the benefit of consumer health, the molecular information contributes more than information from the individual animal or from 30 paternal half-sibs. However, when 60 paternal half-sibs were used the result was the same as Q20.

Table 7.8 *Expected genetic change in each goal trait to improve meat healthiness[‡]*

Response in	Economic value	IND[†]	IND+HS30	IND+Q05	IND+Q10	IND+Q20	IND+Q50
Flavour (units 0-100) **+**	+1	1.55	1.57	1.89	2.19	2.76	4.20
MUFA (mg/100g) (LW+MD+QTL$_{FLAVOUR}$)*	+1	7.31	9.81	8.27	9.22	10.9	15.4
CLA (mg/100g) (LW+MD+QTL$_{CLA}$)*	+1	9.50	10.6	11.21	12.6	15.3	22.2
Oleic acid (mg/100g) (LW+MD+QTL$_{OLEIC}$)*	+1	3.27	3.67	4.87	6.13	8.16	13.0
Flavour (units 0-100) **+**	+1	1.79	1.98	2.00	2.30	2.83	4.27
Linolenic acid (mg/100g) (LW+MD+QTL$_{LINIOLENIC}$)*	+1	2.55	2.83	4.44	5.75	7.86	12.8

[†]The genetic change expressed as % of the mean of the goal trait.
[‡]To derive the genetic progress in each trait for one round of selection, a selection intensity of 1 was assumed.
*Measured ('index') traits to achieve each goal.

The importance of each measured trait in the overall index is shown in Table 7.9. The results show that CT-muscle density is the most important trait for all the four indices, unless a QTL is used, which is then the most important. For example, for index 3 where we assume information only on the individual, if CT-muscle density or the Q10 are dropped from the index, the genetic progress would be reduced by 96% or 47%, respectively.

Table 7.9 *Contribution of each trait to the index*[†]

		IND	IND+HS30	IND+Q05	IND+Q10	IND+Q20	IND+Q50
Index1	**LW +**	0.12	0.06	0.09	0.07	0.04	0.01
	MD (Flavour+MUFA)*	0.70	0.46	0.36	0.22	0.09	0.01
	$QTL_{FLAVOUR}$			0.12	0.21	0.33	0.53
	LW-HS		0.01				
	MD-HS		0.23				
Index2	**LW +**	0.02	0.01	0.03	0.02	0.01	0.01
	MD +	0.95	0.04	0.36	0.21	0.08	0.01
	QTL_{CLA} (CLA)*			0.15	0.25	0.38	0.57
	LW-HS		0.10				
	MD-HS		0.31				
Index3	**LW +**	0.01	0.01	0.06	0.04	0.02	0.01
	MD + (Oleic acid)*	0.96	0.10	0.17	0.07	0.09	0.02
	QTL_{OLEIC}			0.33	0.47	0.60	0.75
	LW-HS		0.01				
	MD-HS		0.32				
Index4	**LW +**	0.01	0.03	0.014	0.013	0.002	0.001
	MD + (Flavour+Linolenic acid)	0.92	0.06	0.15	0.06	0.05	0.02
	$QTL_{LINOLENIC}$			0.35	0.47	0.62	0.76
	LW-HS		0.09				
	MD-HS		0.31				

[†] The higher the value, the more the trait contributes to the index.
*Measured ('index') traits to achieve each goal.

7.3.3.3 Expected benefits from selection for meat shelf-life

Several studies have been undertaken to increase the PUFA content in meat lipids. Ruminants have been fed rations supplemented with protected (from ruminal hydrogenation) or unprotected oilseeds or free oils rich in omega-6 PUFA (sunflower, cotton, canola, safflower, soybean), or in omega-3 PUFA (linseed, fish oils; Clinquart et al., 1995; Demeyer and Doreau, 1999; Wood et al., 1999); however, PUFA are preferential targets for free radical attacks initiating peroxidation. Lipoperoxidation may be involved in the alteration of animal performance (growth, reproduction) and health (immunological disturbances), owing to metabolic disturbances (Aurousseau, 2002) as described in humans (Slater, 1984; Pré, 1991). Lipid oxidation produces free radicals and therefore, may impair the health of animals (Scislowski et al., 2005). In muscle tissues it can promote myoglobin (colour) oxidation and lead to the formation of rancid odours and flavours. Meat with more PUFA may be more oxidisable, but when these PUFA are derived from pasture feeding, such as our population, they are associated with more antioxidant in the form of α-tocopherol, carotenoids and flavonoids (Wood and Enser, 1997), which stabilise the fatty acids and make the meat more desirable (Gatellier *et al.*, 2004; Moloney *et al.*, 2001; Richardson *et al.*, 2004). High levels of PUFA may cause meat shelf-life to shorten. This reduction in shelf-life is the result of lipid oxidation products catalysing the oxidation reactions forming dark brown metmyoglobin, and also because these products cause rancidity in cooked meat (Wood *et al.*, 1999).

Thus, in this section we explored two cases of indices with respect to improve meat shelf-life (through PUFA), or equal emphasis on reducing PUFA and increasing meat flavour. The specific goal traits are PUFA and flavour, with live weight and CT-muscle density on individual and combinations of measures on half-sibs or QTL for flavour, as the measured traits and are shown in Table 7.10. The first case was to decrease PUFA and the second, to simultaneously decrease PUFA and increase flavour of meat.

Table 7.10 *Breeding goals and index (measured) traits to improve meat shelf-life*

Breeding goal traits	Index traits
PUFA	Live weight on individual and half-sibs + CT-muscle density on individual and half-sibs
Flavour + PUFA	Live weight on individual and half-sibs + CT-muscle density on individual and half-sibs + 'known" QTL for flavour explaining 5, 10, 20 or 50% of the goal trait (flavour)

The results of the genetic change from one round of selection, expressed as a proportion of the mean, for each goal trait are shown in Table 7.11. Using measurements such as CT-muscle density and molecular information, we can possibly decrease PUFA and keep the flavour of meat more or less constant. Therefore, as the content of PUFA in the meat decreased by 11.9 and 13.7% of the mean for the index case of using just information from the animal and combination on half-sibs, respectively, the reduction in the flavour score was negligible. In addition, PUFA were positively correlated with rancid flavour, which means that reduction of PUFA will result in reduction in rancidity of 5% and, thus, increase shelf-life. Hence, these results show that improvement of shelf-life of meat is possible, through a small decrease in PUFA, with flavour score remaining unchanged.

Table 7.11 *Expected genetic change in each goal trait to improve meat healthiness*[‡]

Response in	Economic value	IND[†]	IND+ HS30	IND+ Q05	IND+ Q10	IND+ Q20	IND+ Q50
PUFA (mg/100g) (LW+MD)	-1	-11.9	-13.7				
Flavour (units 0-100) **+**	+1	-0.32	-0.33	-0.41	-0.49	-0.63	-1.03
PUFA (mg/100g) (LW+MD+QTL$_{FLAVOUR}$)	-1	-11.8	-13.6	-12.7	-13.6	-15.2	-20.0

[†]The genetic change expressed as % of the mean of the goal trait.
[‡]To derive the genetic progress in each trait for one round of selection, a selection intensity of 1 was assumed.
*Measured ('index') traits to achieve each goal.

The importance of each measured trait in the overall index is shown in Table 7.12. The results show that again CT-muscle density is the most important trait for all the four indices, unless a QTL is used, which is then the most important. For example, in the case of using index 1, to improve shelf-life, decreasing PUFA, if CT-muscle density, measured on individual was dropped from the index, the genetic progress would be reduced by 50%. Hence, improvement in shelf-life of meat can be achieved, decreasing PUFA and at the same time rancidity, and with flavour score remaining constant.

Table 7.12 *Contribution of each trait to the index*[†]

		IND	IND+HS30	IND+Q05	IND+Q10	IND+Q20	IND+Q50
Index1	**LW +**	0.25	0.07				
	MD (PUFA)*	0.50	0.08				
	LW-HS		0.12				
	MD-HS		0.20				
Index2	**LW +**	0.24	0.07	0.19	0.15	0.11	0.04
	MD +	0.52	0.08	0.34	0.24	0.12	0.05
	QTL$_{FLAVOUR}$ (LW+MD+QTL$_{FLAVOUR}$)*			0.07	0.13	0.23	0.42
	LW-HS		0.11				
	MD-HS		0.21				

[†] The higher the value, the more the trait contributes to the index.
*Measured ('index') traits to achieve each goal.

7.4 Changing the way of breeding

The importance of meat quality traits to the sheep meat industry is beyond question. However, their practical improvement from a genetic standpoint is limited by the fact that no industry-wide programme has been established for their improvement. Indirectly, however, through positive genetic associations between growth rate, live weight and muscle area, it appears likely that breeders have been selecting for improved lean meat yield already.

Genetic gains made within breeding flocks accrue to other sheep industry participants through sale of rams to commercial sheep farmers. Most farmers are in business to make a profit, and this can be affected through purchase of rams from breeders which increase income, or decrease expenses (or both). Currently farmers are rewarded for increased carcass weight. This is reflected by the current breeding goals (Simm and Dingwall, 1989), which essentially increase growth rate and lean carcass weight. Predicted correlated responses, using this current index, in meat quality traits (*e.g.* IMF, lamb flavour, PUFA) are found to be trivial. Therefore, the current index is neutral with regard to meat quality.

The motivation for breeding sheep with superior carcass and meat quality characteristics would have to arise through improved ram prices paid by sheep farmers. Farmers would in-turn have to be motivated to pay more for rams from which a high percentage of progeny meet the requirements of processors and attract premium prices. Recent initiatives to improve carcass feedback to farmers may provide motivation to purchase rams with improved carcass and meat quality.

Which of the above traits would be important for improvement in selection programmes? Some traits such as muscle density can be improved relatively efficiently by CT scanning potential breeding animals and incorporation of scanning measures into a genetic evaluation. Although more accurate, CT scanning has disadvantages relative to other scanning means, such as ultrasound scanning. Firstly, CT is expensive. Secondly, CT scanning units are typically situated at fixed locations, requiring transport of animals to and from the CT facility, whereas, for example, ultrasound is readily portable. Therefore, CT is likely to be most economically beneficial to the sheep industry when applied in a two-stage selection programme where initial screening of selection candidates is done using other means (*e.g.* ultrasound) (Jopson *et al.*, 1995 and 1997; Lewis and Simm, 2002) or when CT scanning a small number of relatives, such as 30 half-sibs of fewer, as it was suggested in this **Chapter**. To design a two-stage selection programme of this type requires information on the accuracy of both ultrasound and CT scanning in predicting carcass composition in the breed types to be selected (Jopson *et al.*, 1995).

Other traits such as flavour, juiciness, and meat fatty acids are more difficult to improve through selection. In addition, there are negative genetic correlations between some meat quality traits, for example improving intramuscular fat content can lead to an increased carcass fatness and possibly to decrease in lean meat yield.

This **Chapter** explored the potential benefit of improving meat flavour, meat healthiness and meat shelf-life in sheep by selection. The selection indices used here included information on: i) individual animal, ii) relatives and iii) *"known"* QTL. In all cases the information gained from the use of QTL depended on the heritability of the trait and the genetic correlation with CT-muscle density. In general, the QTL effect had to be quite large to give substantial improvements. As a broad summary, the results demonstrated that using a QTL which explained more than 20% of the genetic variance of the trait, the genetic progress was 20-30% higher than using phenotypic information on the individual or its relatives.

The question arising is which direction one wants to follow (Table 7.13).

Table 7.13 *Implications when selection for and against certain meat quality traits*

Supposed breeding goal	Result of selection
Improved meat flavour	✓ Higher intramuscular fat ✓ Better flavour ✓ Juicier meat
Improved meat healthiness	✓ Increase in total MUFA, CLA, Oleic acid ✓ Decrease in total SFA ✓ Better flavour
Improved meat shelf-life (Decrease PUFA content)	✓ Decrease total PUFA ✓ Decreased rancidity ✓ Unchanged flavour

From the perspective of consumers, the demand is consistent for meat quality and nutritional value, with increasing MUFA and PUFA, and reducing the amount of SFA in sheep meat. On the other hand, the retailers demand increasing shelf-life of the final product, and although we have showed that this can be achieved by selection, it may decrease the amount of PUFA and flavour score.

7.5 Perspectives, conclusions and recommendations

This thesis has raised interesting questions which were explored in each chapter, but some new concerns brought forward which will involve further research. The first aspect of study to further these investigations should be extended to more QTL studies. It is important to determine if the same QTL identified in this study are segregating in other flocks, and particularly commercial ones, which are under selection for meat quality and production traits.

To expand upon the results presented in **Chapters 4 and 5,** it would be interesting to complete the genome scan, as a partial scan was carried out – only eight chromosomes were genotyped due to financial constraints. Given the quality of the results obtained, completing the genome scan may yield yet more interesting results. Furthermore, it would be of great interest to investigate the eight QTL for fatty acids identified in the same location with large effects on chromosome 21 (**Chapter 5**) that probably would correspond to a single QTL. Hence, the next step would require more precise genetic markers. Also, several genes might be selected as candidate genes to explain the fatty acid QTL identified in **Chapter 5**, as fatty acid metabolism is influenced by large numbers of genes involved in complex metabolic routes, and these issues should be under further research, as well. Association analyses between allelic variants of these genes and fatty acid content would need to be performed to find the necessary genetic markers.

Consider the results for CT-muscle density presented in **Chapters 3 and 6**, CT-muscle density appeared to influence important aspects of meat quality, such as intramuscular fat, flavour, juiciness, and fatty acid content. Thus, muscle density could be used in selection strategies that attempt to simultaneously improve lean meat and meat characteristics. This was discussed in **Section 7.3**, where different selection procedures explored the improvement of meat quality with CT-muscle density as a measured trait. Moreover, to our knowledge CT-muscle density is now being explored in sheep breeding programmes in New Zealand (Campbell and Waldron, 2006), as a direct result of the findings of this thesis.

This thesis has identified new information on the genetic basis of fatty acids in sheep meat and new *in vivo* predictors of meat quality in sheep. A trial being done in Merino sheep, in Australia, by Greef *et al.* (2006), confirmed that fatty acid content of meat is indeed heritable. This suggests that composition of lamb meat can be changed genetically and the next step of this project (personal communication with S. Kitessa, CSIRO) is to include those fatty acids in a traditional breeding programme.

Regarding other species such as beef cattle, there is the "RoBoGen" project, which is performed by Roslin Institute, and they have identified many QTL affecting meat quality with a complete genome scan. However, the results remain confidential. In pigs, most of the studies in identifying QTL for meat quality and fatty acids are focused in subcutaneous fat rather than intramuscular fat (Perez-Ensizo *et al.*, 2000; Clop *et al.*, 2003). A recent example for the importance of fatty acid content of meat for human health was the creation of transgenic pigs rich in omega-3 fatty acids (Lai *et al.*, 2006), although consumers are likely to reject such technology, because of the negative perceived effects on animal welfare. Also, in pigs known major genes affecting meat quality have been identified, such as the halothane- and the RN-gene (Naveau *et al.*, 1985; Le Roy *et al.*, 1990; Fujii *et al.*, 1991), which have a causal negative effect on meat quality traits. The IMF-gene is believed to optimise the eating quality and tenderness of pork (Hovenier, 1993).

The trend towards healthy (low fat) meat products has changed the demands of the consumer. Not only is breeding about the future, but the end product of breeding (lamb meat) is moving continuously to the forefront to the consumer. Hence, genetics is the bullet on it's way (through the supply chain) to the target (a satisfied consumer). This thesis illustrates how meat quality of lamb meat can and must be changed in order to satisfy consumer's needs. Thus, it is time to restructure the breeding goals in the sheep sector for a more consumer orientated production.

In summary, genetic variation of meat quality traits in sheep has been quantified. Furthermore, the incorporation of *in vivo* and molecular information into a selection index for improving the genetic profile of fatty acid in sheep meat has been developed in this thesis. It has been shown, for the first time in sheep, that the use of CT-muscle density may be a means of making broad improvements in the perceived and actual quality of sheep meat. The key finding of this thesis is the prediction of fatty acid composition *in vivo*. Moreover, if identifying candidate genes to eventually predict the fatty acid composition of lamb muscle tissue based on DNA analysis is successful, this would lead to the development of lamb with an improved fatty acid profile of "heart-healthy" lamb meat, without asking consumers to drastically change food choices and without compromising the health of the sheep or the flavour of the meat.

APPENDIX II

Table 7.14 *Accuracy of selection index to improve meat flavour*

Breeding goal	IND	IND+HS30	IND+Q05	IND+Q10	IND+Q20	IND+Q50
IMF (LW+MD)	0.47	0.53				
IMF + Flavour (LW+MD+QTL$_{FLAVOUR}$)	0.47	0.54	0.50	0.53	0.58	0.74
IMF + Juiciness (LW+MD)	0.47	0.54				
Flavour (LW+MD+QTL$_{FLAVOUR}$)	0.30	0.33	0.35	0.40	0.48	0.71

Table 7.15 *Accuracy of selection index to improve meat healthiness*

Breeding goal	IND	IND+HS30	IND+Q05	IND+Q10	IND+Q20	IND+Q50
Flavour + MUFA (LW+MD+QTL$_{FLAVOUR}$)	0.34	0.45	0.38	0.43	0.51	0.73
CLA (LW+MD+QTL$_{CLA}$)	0.30	0.34	0.35	0.40	0.49	0.71
Oleic acid (LW+MD+QTL$_{OLEIC}$)	0.18	0.20	0.27	0.34	0.45	0.72
Flavour + Linolenic acid (LW+MD+QTL$_{LINOLENIC}$)	0.22	0.24	0.33	0.42	0.56	0.90

Table 7.16 *Accuracy of selection index to improve meat shelf-life*

Breeding goal	IND	IND+HS30	IND+Q05	IND+Q10	IND+Q20	IND+Q50
PUFA (LW+MD)	0.44	0.51				
Flavour + PUFA (LW+MD+QTL$_{FLAVOUR}$)	0.42	0.50	0.44	0.48	0.54	0.70

Bibliography

Aalhus, J. L., Jones, S. D. M., Robertson, W. M., Tong, A. K. W. and Sather, A. P. 1991. Growth characteristics and carcass composition of pigs with known genotypes for stress susceptibility over a weight range of 70 to 120 kg. *Animal Production* **52**: 347-353.

Aaslyng, M. D., Bejerholm, C., Ertbjerg, P., Bertram, H. C. and Andersen, H. J. 2003. Cooking loss and juiciness of pork in relation to raw meat quality and cooking procedure. *Food Quality and Preference,* **14:** 277-288.

Abdullah, A. Y., Purchas, R. W., and Davies, A.S. 1998. Patterns of change with growth for muscularity and other composition characteristics of Southdown rams selected for high and low backfat depth. *New Zealand Journal of Agricultural Research* **41**: 367-376.

Addis, P. B., and Park, S. W. 1989. Role of lipid oxidation products in atherosclerosis. In: S. Raylor and R. Scanlan (Ed.) Food Toxicology, *A Perspective on the Relative Risks*, p297. Marcel Dekker, New York.

Afonso, J. and Thompson, J. M. 1996. Fat distribution in sheep selected for/against backfat depth, during growth on ad libitum feeding. *Livestock Production Science* **46:** 97-106.

Allen, P. 1990. Reducing fat in meat animals. In *New approaches to measuring body composition in live animals* (ed. J. D. Wood and A. V. Fisher), pp. 201-254. Elsevier, London, UK.

Alliston, J. C. 1983. Evaluation of carcass quality in the live animal. In *Sheep production* (ed. W. Haresign), pp. 75-95. Butterworths, London.

American Heart Association. 1999 *Heart and stroke statistical update.* Dallas, Texas: American Heart Association, 1998.

Ashgar, A., Gray, J. I., Buckley, D. J., Pearson, A. M., and Booren, A. M. 1988. Perspectives on warmed-over flavor. *Food Technology* **42**: 102-108.

Association of Official Analytical Chemists. 1997. *Official methods of analysis, 16th edition.* AOAC, Arlington, VA.

Aurousseau, B. 2002. Les radicaux libres dans l'organisme des animaux d'elevage: Consequences sur la reproduction, la physiologie et la qualite de leurs produits. *INRA Productions Animales* **15:** 67–82.

Avery, N. C., Sims, T. J., Warkup, C. and Bailey, A. J. 1996. Collagen cross-linking in porcine m. longissimus lumborum: Absence of a relationship with variation in texture at pork weight. *Meat Science* **42:** 355-369.

Bari, F., Khalid, M., Haresign, W., Merrell, B., Murray, A. and Richards, R. I. W. 1999. An evaluation of the success of MOET in two breeds of hill sheep maintained under normal systems of hill flock management. *Animal Science* **69:** 367-376.

Bari, F., Khalid, M., Haresign, W., Murray, A. and Merrell, B. 2000. Effect of mating system, flushing procedure, progesterone dose and donor ewe age on the yield and quality of embryos within a MOET program in sheep. *Theriogenology* **53:** 727-742.

Barkhouse, K. L., Van Vleck, L. D., Cundiff, L. V., Koohmaraie, M., Lunstra, D. D. and Crouse, J. D. 1996. Prediction of breeding values for tenderness of market animals from measurements on bulls. *Journal of Animal Science* **74:** 2612-2621.

Bauchart, D., Gruffat D. and Durand D. 1996. Lipid absorption and hepatic metabolism in ruminants. *Proceedings of the Nutritional Society* **55:** 39–47.

Bauchart, D., Vérité R. and Rémond B. 1984. Long-chain fatty acid digestion in lactating cows fed fresh grass from spring to autumn. *Canadian Journal of Animal Science* **64:** 330–331.

Baulain, U. 1997. Magnetic resonance imaging for the in vivo determination of body composition in animal science. *Computers and Electronics in Agriculture* **17:** 189-203.

Bejerholm, C. and Barton-Gade, P. 1986. Effect of intramuscular fat level on eating quality of pig meat. In *32nd European meeting of meat research workers,* Ghent, Belgium. 389-391.

Bendall, J. R. 1951. The shortening of rabbit muscles during rigor mortis: its relation to the breakdown of adenosine triphosphate and creatinphosphate and to muscular contraction. *Journal of Physiology* **114:** 71-88.

Bendall, J. R. 1973. Post mortem changes in muscles. In G. H. Bourne, *Structure and function of muscle,* pp243-309. Academic Press. New York.

Bendall, J. R. and Swatland, H. J. 1988. A review of the relationships of pH with physical aspects of pork quality. *Meat Science* **24:** 85-126.

Bendall, J. R. and Wismer-Pedersen, J. 1962. Some properties of the fibrillar proteins of normal and watery pork muscle. *Journal of Food Science* **27:** 144-159.

Bennett, G. L., Johnson, D. L., Kirton, A. H. and Carter, A. H. 1991. Genetic and environmental effects on carcass characteristics of Southdown x Romney lambs: II. Genetic and phenotypic variation. *Journal of Animal Science* **69:** 1864-1874.

Bishop, S. C. 1993. Selection for predicted carcass lean content in Scottish Blackface sheep. *Animal Production* **56:** 379-386.

Bishop, S. C. 1994. Genetic relationships between predicted and dissected carcass composition in Scottish Blackface sheep. *Animal Production* **59:** 321-478.

Bishop, S. C., Cameron, N. D., Speake, B. K., Bracken, J., Ratchford, I. A. J. and Noble, R. C. 1995. Responses in adipocyte dimensions to divergent selection for predicted carcass lean content in sheep. *Animal Science* **60:** 215-221.

Bishop, S. C., Conington, J., Waterhouse, A. and Simm, G. 1996. Genotype x environment interactions for early growth and ultrasonic measurements in hill sheep. *Animal Science* **62:** 271-277.

Boccard, R., Buchter, L., Casteels, E., Cosentino, E., Dransfield, E., Hood, D. E., Joseph, R. L., MacDougall, D. B., Rhodes, D. N. and Schon, I. 1981. Procedures for measuring meat quality characteristics in beef production experiments. Report of a working group in the commission of the European communities' (CEC) beef production research programme. *Livestock Production Science* **8:** 385-397.

Bonanome, A. and Grundy, S. M. 1988. Effect of dietary stearic-acid on plasma cholesterol and lipoprotein levels. *New England Journal of Medicine* **318:** 1244-1248.

Bovenhuis, H., van Arendonk, J. A. M., Davis, G., Elsen, J.-M., Haley, C. S., Hill, W. G., Baret, P. V., Hetzel, D. J. S. and Nicholas, F. W. 1997. Detection and mapping of quantitative trait loci in farm animals. *Livestock Production Science* **52:** 135-144.

Bradford, G. E. and Spurlock, G. M. 1972. Selection for meat production in sheep - Results of a progeny test. *Journal of Animal Science* **34:** 737-745.

Briskey, E. J. 1964. Etiological status and asociated studies of pale, soft, exudative porcine musculature. In C. O. Chichester, E. M. Mrak and G. F. Stewart, *Advances in Food Research,* pp90-168. Academic Press. London.

Broad, T. E., Glass, B. C., Greer, G. J., Robertson, T. M., Bain, W. E., Lord, E. A. and McEwan, J. C. 2000. Search for a locus near to myostatin that increases muscling in

Texel sheep in New Zealand. *Proceedings of New Zealand Society of Animal Production* **60:** 110-112.

Buckley, D. J., Gray, J. I., Asghar, A., Booren, A. M. , Crackel, R. L., Price, J. F. and Miller, E. R. 1989. Effects of dietary antioxidants and oxidized oil on membranal lipid stability and pork product quality. *Journal of Food Science* **54:**1193.

Butterfield, R. M. 1988. *New concepts of sheep growth.* University of Sydney: Sydney. 168p.

Cameron, N. D. 1990. Genetic and phenotypic parameters for carcass traits, meat and eating quality traits in pigs. *Livestock Production Science* **26:** 119-135.

Cameron, N. D. and Bracken, J. 1992. Selection for carcass lean content in a terminal sire breed of sheep. *Animal Production* **54:** 367-377.

Cameron, N. D. and M. B. Enser. 1991. Fatty acid composition of lipid in longissimus dorsi muscle of Duroc and British Landrace pigs and its relationship with eating quality. *Meat Science* **29:** 295-307.

Cameron, N. D., Bishop, S. C., Speake, B. K., Bracken, J. and Noble,R. C. 1994. Lipid composition and metabolism of subcutaneous fat in sheep divergently selected for carcass lean content. *Animal Production* **58:** 237-242.

Cameron, N. D., Enser, M., Nute, G. R., Whittington, F. M., Penman, J. C., Fisken, A. C., Perry, A. M. and J. D. Wood. 2000. Genotype with nutrition interaction on fatty acid composition of intramuscular fat and the relationship with flavour of pig meat. *Meat Science* **55:** 187-195.

Campbell, A. W. and Waldron, D. F. 2006. **Genetic improvement of meat production in small ruminants** 2006. *Proceedings of the 8th World Congress on Genetics Applied to Livestock Production,* Communication no. 04-06.

Campbell, A. W., Bain, W. E., McRae, A. F., Broad, T. E., Johnstone, P. D., Dodds, K. G., Veenvliet, B. A., Greer, G. J., Glass, B. C. and Beattie, A. E. 2003. Bone density in sheep: genetic variation and quantitative trait loci localisation. *Bone* **33:** 540-548.

Casas, E., R. T. Stone, J. W. Keele, S. D. Shackelford, S. M. Kappes and Koohmaraie, M. 2001. A comprehensive search for QTL affecting growth and carcass composition of cattle segregating alternative forms of the myostatin gene. *Journal of Animal Science* **79:** 854–860.

Casas, E., Shackelford, S. D., Keele, J. W., Stone, R. T., Kappes, S. M. and Koohmaraie, M. 2000. Quantitative trait loci affecting growth and carcass composition of cattle segregating alternate forms of myostatin. *Journal of Animal Science* **78:** 560-569.

Casas-Carrillo, E., Kirkpatrick, B. W., Prill-Adams, A., Price, S. G. and Clutter, A. C. 1997. Relationship of growth hormone and insulin-like growth factor-1 genotypes with growth and carcass traits in swine. *Animal Genetics* **28:** 88-93.

Churchill, G. A. and Doerge, R. W. 1994. Empirical threshold values for quantitative trait mapping. *Genetics* **138:** 963-971.

Claus, R., Weiler, U. and Herzog, A. 1994. Physiological aspects of androstenone and skatole formation in the boar-A review with experimental data. *Meat Science* **38:** 289-305.

Clinquart, A., Micol, D., Brundseaux, C., Dufrasne, I. and Istasse, L. 1995. Utilisation des matières grasses chez les bovines à l'ebgraissement. *INRA Productions Animales* **8:** 29–42.

Clop, A., Marcq, F., Takeda, H., Pirottin, D., Tordoir, X., Bibe, B., Bouix, J., Caiment, F., Elsen, J. M., Eychenne, F., Larzul, C., Laville, E. , Meish, F., Milenkovic, D., J. Tobin, J., Charlier, C. and Georges, M. 2006. A mutation creating a potential illegitimate microRNA target site in the myostatin gene affects muscularity in sheep. *Nature Genetics* **38:** 813-818.

Clop, A., Ovilo, C., Perez-Enciso, M., Cercos, A., Tomas, A., Fernandez, A., Coll, A., Folch, J. M., Barragan, C., Diaz, I., Oliver, M. A., Varona, L., Silio, L., Sanchez, A. and L. L. Noguera. 2003. Detection of QTL affecting fatty acid composition in the pig. *Mammalian Genome* **14:** 650-656.

Clutter, A. C. 1995. *Molecular genetics and meat quality.* National Swine Improvement Federation, Iowa.

Cockett, N. E., Jackson, S. P., Shay, T. L., Nielsen, D., Moore, S. S., Steele, M. R., Barendse, W., Green, R. D. and Georges, M. 1994. Chromosomal localization of the callipyge gene in sheep using bovine DNA markers. *Proceedings of the National Academy of Sciences* **91:** 3019-3023, USA.

Cockett, N. E., Jackson, S., Shay, T. L., Farnir, F., Berghmans, S., Snowder, G., Nielsen, D. and Georges, M. 1996. Polar overdominance at the ovine callipyge locus. *Science* **273:** 236-238.

Collins, A. C., Martin, I. C. A. and Kirkpatrick, B. W. 1993. Growth Quantitative Trait Loci (Qtl) on Mouse Chromosome-10 in A Quackenbush-Swiss X C57Bl/6J Backcross. *Mammalian Genome* **4:** 454-458.

Conington, J., Bishop, S. C., Grundy, B., Waterhouse, A. and Simm, G. 2001. Multitrait selection indexes for sustainable UK hill sheep production. *Animal Science* **73:** 413-423.

Conington, J., Bishop, S. C., Waterhouse, A. and Simm, G. 1995. A genetic-analysis of early growth and ultrasonic measurements in hill sheep. *Animal Science* **61:** 85-93.

Conington, J., Bishop, S. C., Waterhouse, A. and Simm, G. 1998. A comparison of growth and carcass traits in Scottish Blackface lambs sired by genetically lean or fat rams. *Animal Science* **67:** 299-309.

Cook, M. E., Miller, C. C., Park, Y. and Pariza, M. 1993. Immune modulation by altered nutrient metabolism - Nutritional control of immune-induced growth depression. *Poultry Science* **72:** 1301-1305.

Cramer, D. A., Pruett, J. B., Kattnig, R. M. and Schwartz W. C. 1970a. Comparing breeds of sheep. I. Flavor differences. *Proceedings of the Western Section, American Society of Animal Science* **21:** 267–269.

Cramer, D. A., Pruett, J. B., Swanson, V. B., Schwartz, W. C., Kattnig, R. M., Phillips, B. L. and Wookey, L. E. 1970b. Comparing breeds of sheep. II. Carcass Characteristics. *Proceedings of the Western Section, American Society of Animal Science* **21:** 270–272.

Crews, D. H., Jr., Pollak, E. J., Weaber, R. L., Quaas, R. L. and Lipsey, R. J. 2003. Genetic parameters for carcass traits and their live animal indicators in Simmental cattle. *Journal of Animal Science* **81:** 1427-1433.

Crouse, J. D., Busboom, J. R., Field, R. A. and Ferrell, C. L. 1981. The effect of breed, diet, sex, location and slaughter weight on lamb growth, carcass composition and meat flavour. *Journal of Animal Science* **53:** 376-387.

Cummings, S. R., Black, D. M., Nevitt, M. C., Browner, W. S., Cauley, J. A., Genant, H. K., Mascioli, S. R., Scott, J. C., Seeley, D. G., Steiger, P. and Vogt, T. M. 1990. Appendicular Bone-Density and Age Predict Hip Fracture in Women. *Jama-Journal of the American Medical Association* **263:** 665-668.

Cummings, S. R., Black, D. M., Nevitt, M. C., Browner, W. S., Cauley, J. A., Genant, H. K., Mascioli, S. R., Scott, J. C., Seeley, D. G., Steiger, P. and Vogt, T. M. 1990.

Appendicular Bone-density and age predict hip fracture in women. *Journal of the American Medical Association* **263**: 665-668.

Cunningham, E. P. 1969. *Animal breeding theory.* Internordic licentiat course notes in quantitative genetics, Norway.

Cunningham, E. P. 1975. Multi-stage index selection. *Theoretical and Applied Genetics* **46**: 55-61.

Davis, G. P., Hetzel, D. J. S., Corbet, N. J., Scacheri, S., Lowden, S., Renaud, J., Mayne, C., Stevenson, R., Moore, S. S. and Byrne, K. 1998. The mapping of quantitative trait loci for birth weight in a tropical beef herd. *Proceedings of the 6th World Congress on Genetics Applied to Livestock Production* **26**: 441-444.

De Boer, H. D., Dumont, B. L., Pomeroy, R. W. and Weniger, J. H. 1974. Manual on E.A.A.P. Reference Methods for the assessment of carcass characteristics in cattle. *Livestock Production Science* **1**: 151-164.

De Koning, D. J., Schulmant, N. F., Elo, K., Moisio, S., Kinos, R., Vilkki, J. and Maki-Tanila, A. 2001. Mapping of multiple quantitative trait loci by simple regression in half-sib designs. *Journal of Animal Science* **79**: 616-622.

De Lorgeril M., Renaud S. and Mamelle N. 1994. Mediterranean alpha linolenic acid-rich diet in secondary prevention of coronary heart disease, *Lancet* **343**: 1454 1459.

De Smet, S., Raes, K. and D. Demeyer. 2004. Meat fatty acid composition as affected by fatness and genetic factors: a review. *Animal Research* **53**: 81-98.

Demeyer, D. and Doreau, M. 1999. Targets and procedures for altering ruminant meat and milk lipids. *Proceedings of the Nutrition Society* **58**: 593–607.

Demirel, G., Wachira, A. M., Sinclair, L. A., Wilkinson, R. G., Wood, J. D. and Enser, M. 2004. Effects of dietary n-3 polyunsaturated fatty acids, breed and dietary vitamin E on the fatty acids of lamb muscle, liver and adipose tissue. *British Journal of Nutrition* **91**: 551-565.

Denke, M. A. 1994. Role of beef and beef tallow. An enriched source of stearic-acid, in a cholesterol-lowering diet. *American Journal of Clinical Nutrition* **60** (Suppl.): 1044-1049.

Department of Health. 1994. Nutritional Aspects of Cardiovascular Disease, London: Her Majesty's Stationary Office.

DeVol, D. L., McKeith, F. K., Bechtel, P. J., Novakofski, J., Shanks, R. D. and Carr, T. R. 1988. Variation in composition and palatability traits and relationships between muscle characteristics and palatability in a random sample of pork carcasses. *Journal of Animal Science* **66**: 385.

Dikerman, M. E. 1994. Genetics of meat quality. In: P*roceedings of the 5th World Congress on Genetics Applied to Livestock Production* **19**: 437-438.

Dransfield, E., Nute, G. R., MacDougall, D. B. and Rhodes, D. N. 1979. Effect of sire breed on eating quality of crossbred lambs. *Journal of Science Food and Agriculture* **3**: 805–808.

Dransfield, E., Mottram, G. R., Rowan, T. G. and Lawrence, T. L. J. 1985. Pork quality from pigs fed on low glucosinate rapeseed meal: influence of level in the diet, sex, ultimate pH. *Journal of Science Food and Agriculture* **36**: 546-556.

Duckett, S. K., Cuvala, S. L. and Snowder, G. D. 1999. Effects of Dorper genetics on tenderness, fatty acid and cholesterol content of lamb. *Journal of Animal Science* **77** (Suppl. 1): 168.

Duckett, S. K., Snowder, G. D. and Cockett, N. E. 2000. Effect of the callipyge gene on muscle growth, calpastatin activity, and tenderness of three muscles across the growth curve. *Journal of Animal Science* **78**: 2836-2841.

Dumont, B. L. 1957. Nouvelles methodes pour l'estimation de la qualite? des carcasses sur les porcs vivants [New methods of estimation of carcass quality of live pigs]. *Report of the Meeting on Pig Progeny Testing in Europe* 23.

Duniec, H. 1961. Heritability of chemical fat content in the loin muscle of baconers. *Animal Production* **3**: 195.

East, E., Taylor, J., Miller, I. T. and Widdowson, R. W. 1959. Measurements of back fat thickness on live pigs by ultrasonics. *Animal Production* **1**: 129-134.

Eikelenboom, G., Hoving-Bolink, A. H. and van der Wal, P. G. 1996. The eating quality of pork. 2. The influence of intramuscular fat. *Fleischwirtschaft* **76**: 517–518.

Elmore, J. S., Mottram, D. S., Enser, M. and Wood, J. D. 2000. The effects of diet and breed on the volatile compounds of cooked lamb. *Meat Science* **55**: 149–159.

Elo, K. T., Vilkki, H. J., De Koning, D. J., Velmala, R. J. and Maki-Tanila, A. V. 1999. A quantitative trait locus for live weight maps to bovine chromosome 23. *Mammalian Genome* **10**: 831-835.

Emmans, G. C., Kyriazakis, I. and Fisher, C. 2000. Consequences of selecting for growth and body composition characteristics in poultry and pigs. p39-53 *The challenge of* *genetic change in animal production.* Ed. Hill, W.G.; Bishop, S.C.; McGuirk, B.; McKay, J.C.; Simm, G.; Webb, A.J. *British Society of Animal Science.* Occasional Publication No. 27.

Enfält, A. C., Lundström, K., Hansson, I., Johansen, S. and Nyström, P. E. 1997a. Comparison of noncarriers and heterozygous carriers of the RN allele for carcass composition, muscle distribution and technological meat quality in Hampshire-sired pigs. *Livestock Production Science* **47:** 221-229.

Enser M., Scollan N. D., Choi N. J., Kurt E., Hallett K. and Wood J. D. 1999. Effect of dietary lipid on the content of conjugated linoleic acid in beef muscle, *Journal of Animal Science* **69:** 43–146.

Enser, M. 2001. The role of fats in human nutrition. In *Oils and fats, animal carcass fats,* ed. B. Rossell, pp.77-123. Leatherhead Publishing, Leatherhead, England.

Enser, M., Hallet, K, Hewett, B., Fursey, G. A. J. and Wood, J. D. 1996. Fatty acid content and composition of English beef, lamb, pork at retail. *Meat Science* **44:** 443-458.

Essén-Gustavsson, B., Karlsson, A., Lundström, K. and Enfält, A.-C. 1994. Intramuscular fat and muscle fibre lipid contents in halothane-gene-free pigs fed high or low protein diets and its relation to meat quality. *Meat Science* **38:** 269-277.

Estrade, M., Vignon, X. and Monin, G. 1993a. Effect of the RN gene on ultrastructure and protein fractions in pig muscle. *Meat Science* **35:** 313-319.

Estrade, M., Vignon, X., Rock, E. and Monin, G. 1993b. Glycogen hyperaccumulation in white muscle fibres of RN gene carrier pigs. A biochemical and ultrastructural study. *Comparative Biochemistry and Physiology* **104B:** 321 326.

Falconer, D. S. 1981. Introduction to quantitative genetics. 2nd ed. Longman, London.

Fennessy, P. F., Greer, G. J., Bain, W. E. and Johnstone, P. D. 1993. Progeny test of ram lambs selected for low ultrasonic backfat thickness or high post-weaning growth rate. *Livestock Production Science* **33:** 105-118.

Fernandes, T. L., Wilton, J. W., Mandell, I. B. and Devitt, C. J. B. 2002. Genetic parameter estimates for meat quality traits in beef cattle managed under a constant finishing program. *Proceedings of the 7th World Congress on Genetics Applied to Livestock Production,* Communication no 2-93.

Fernandez, A., de Pedro, E., Nunez, N., Silio, L., Garcia-Casco, J. and C. Rodriguez. 2003. Genetic parameters for meat and fat quality and carcass composition traits in Iberian pigs. *Meat Science* **64:** 405-410.

Fernandez, X., Monin, G., Talmant, A., Mourot, J. and Lebret, B. 1999. Influence of intramuscular fat content on the quality of pig meat - Composition of the lipid fraction and sensory characteristics of m. longissimus lumborum. *Meat Science* **53:** 59-65.

Fisher, A.V., Enser, M., Richardson, R. I., Wood, J. D., Nute, G. R., Kurt, E., Sinclair, L. A. and Wilkinson, R. G. 2000. Fatty acid composition and eating quality of lamb types derived from four diverse breed x production systems. *Meat Science* **55:** 141-147.

Fogarty, N. M. 1995. Genetic parameters for live weight, fat and muscle measurements, wool production and reproduction in sheep: a review. *Animal Breeding Abstracts* **63:** 101-143.

Fogarty, N. M., Safari, E., Taylor, P. J. and Murray, W. 2003. Genetic parameters for meat quality and carcass traits and their correlation with wool traits in Australian Merino sheep. *Australian Journal of Agricultural Research* **54:** 715-722.

Folch, J., Lees, M. and Stanley, G.H.S. 1957. A simple method for the isolation and purification of lipids from animal tissues. *Journal of Biological Chemistry* **226:** 497-509.

Fowler, P. A., Fuller, M. F., Glasbey, C. A., Cameron, G. G. and Foster, M. A. 1992. Validation of the in vivo measurement of adipose tissue by magnetic resonance imaging of lean and obese pigs. *American Journal of Clinical Nutrition* **56:** 7-13.

Fox, C. W., McArthur, J. A. B. and Sather, L. 1962. Effect of sire and breed on flavor scores from weanling lamb. *Journal of Animal Science* **21:** 665 (Abstr.).

Fox, C. W., Eller, R., Sather, L. and McArthur, J. A. B. 1964. Effects of sire and breed on eating qualities from weanling lambs. *Journal of Animal Science* **23:** 596(Abstr.).

Freking, B. A., Keele, J. W., Shackelford, S. D., Wheeler, T. L., Koohmaraie, M., Nielsen, M. K. and Leymaster, K. A. 1999. Evaluation of the ovine callipyge locus: III. Genotypic effects on meat quality traits. *Journal of Animal Science* **77:** 2336-2344.

Freking, B. A., Murphy, S. K., Wylie, A. A., Rhodes, S. J., Keele, J. W., Leymaster, K. A., Jirtle, R. L. and Smith, T. P. L. 2002. Identification of the single base change causing the callipyge muscle hypertrophy phenotype, the only known example of polar overdominance in mammals. *Genome Research* **12:** 1496-1506.

Fujii, J., Otzu, K., Zorazo, F., DeLeon, S., Khanna, V. K., Weiler, J., O´Brien, P. J. and MacLennan, D. H. 1991. Identification of a mutation in porcine ryanodine receptor associated with malignant hyperthermia. *Science* **253**: 448-451.

Gariepy, C., Amiot, J. and Nadai, S. 1989. Ante-mortem detection of PSE and DFD by infrared thermography of pigs before stunning. *Meat Science* **25**: 37-41.

Gatellier, P., Mercier, Y. and Renerre, M. 2004. Effect of diet finishing mode (pasture or mixed diet) on antioxidant status of Charolais bovine meat. *Meat Science* **67**: 385-394.

GenStat 7 Committee 2003. *Reference manual.* Oxford University Press, Oxford.

Gilmour, A. R., Cullis, B. R., Welham, S. J. and Thompon, R. 2004. *ASREML: program user manual.* NSW, Agriculture, Orange, Australia.

Govindarajan, S. 1973. Fresh meat colour. *CRC Critical reviews in Food Technology* **1**: 117-140.

Gray, J. I. and Pearson, A. M. 1987. Rancidity and warmed-over flavour. In *Advances in meat research* (ed. A.M., Pearson, and T.R., Dutson), pp. 221-270. Elsevier, Amsterdam.

Gray, J. I. and Pearson, A. M. 1994. Lipid-derived off-flavours in meat formation and inhibition. In *Flavour of meat and meat products* (ed. F. Shahidi), pp. 116-143. Blackie Academic, London.

Gray, J. I., Gomaa, E. A., and Buckley, D. J. 1996. Oxidative quality and shelf life of meats. *Meat Science* **43**: 111-123.

Greeff, J. C., Young, P., Kitessa, S. and Dowling, M. 2006. Preliminary heritability estimates of individual fatty acids in sheep meat. *Australian Society of Animal Production 26th Biennial Conference 2006 Short Communication number*

Green, P., Falls, K. and S. Crooks. 1990. Cri-map version 2.4. Washington University School Medicine, St. Louis, MO, USA.

Grundy, S. M. 1994. Influence of stearic-acid on cholesterol-metabolism relative to other long-chain fatty-acids. *American Journal of Clinical Nutrition* 60 (Suppl.): 986-990.

Grunert, K. G., Harmsen, H., Larsen, H. H., Sørensen, E. and Bisp, S. 1997. New areas in agricultural and food marketing. In: *Agricultural Marketing and Consumer Behaviour in a Changing World*, pp. 3-30, J.-B. Steenkamp,A. van Tilburg,B. Wierenga,K. G. Grunert and M. Wedel (Ed. Eds.). Boston, MA: Kluwer Academic Publishers.

Haley, C. 2001. Mapping genes for milk and meat quality. *Proceedings of the British Society of Animal Science.*

Hammond, A. C., Carlson, J. R. and Willet, J. R. 1979. The metabolism and disposition of 3-methylindole in goats. *Life Sciences* **25**: 1301-1306.

Hawkins, R. R., Kemp, J. D., Ely, D. G., Fox, J. D., Moody, W. G. and Vimini, R. J. 1985. Carcass and meat characteristics of crossbred lambs born to ewes of different genetic types and slaughtered at different weights. *Livestock Production Science* **12**: 241–250.

Heelsum, A. M. v., Lewis, R. M., Jones, D. W., Haresign, W. and Davies, M. H. Genetic parameters for live weight, ultrasonic measurements and conformation in Bluefaced Leicester sheep. *Proceedings of the British Society of Animal Science,* 122.

Hennig, M. D. 1992. Magnetic resonance imaging for the assessment of body composition in pigs. *Pig News Info* **13**.

Henniñgsson, T. and Malmfors, G. The relationship between live weight of Swedish lambs at 120 days of age and carcass traits. *Proceedings of the Scandinavian Association of Agricultural Scientists,* 256.

Herbein J. H., Loor, J. J. and Wark, W. A. 2000. Conjugated linoleic acids - An opportunity for pasture-based dairy farms? Pages 16-20 in *Proceedings of Mid-Atlantic Dairy Grazing Field Day,* Abingdon, VA.

Hermesch, S., Luxford, B. G. and Graser, H. U. 2000. Genetic parameters for lean meat yield, meat quality, reproduction and feed efficiency traits for Australian pigs: 3. Genetic parameters for reproduction traits and genetic correlations with production, carcase and meat quality traits. *Livestock Production Science* **65**: 261-270.

Hofmann, K. 1994. What is quality? *Meat Focus International* **3**, pp.73–82.

Honikel, K. O. 1998. Reference methods for the assessment of physical characteristics of meat. *Meat Science* **49**: 447-457.

Hopkins, D. L. and Fogarty, N. M. 1998. Diverse lamb genotypes-2. Meat pH, colour and tenderness. *Meat Science* **49**: 477-488.

Hovenier, R. 1993. Breeding for meat quality in pigs. *PhD Thesis.* Department of Animal Breeding, Wageningen Agricultural University, Wageningen, The Netherlands.

Immonen, K., Ruusunen, M. and Puollane, E. 2000. Some effects of residual glycogen concentration on the physical and sensory quality of normal pH beef. *Meat Science* **55**: 33-38.

Ip, C., Singh, M., Thompson, H. J. and Scimeca, J. A. 1994. Conjugated linoleic-acid suppresses mammary carcinogenesis and proliferative activity of the mammary-gland in the rat. *Cancer Research* **54**:1212-1215.

Jacobson, M. and Koehler, H. H. 1963. Components of the flavor of lamb. *Journal of Agricultural and Food Chemistry* **11**:336–339.

Jaime, I., Beltran, J. A., Cena, P. and Roncales, P. 1993. Rapid chilling of light lamb carcasses results in meat as tender as that obtained using conventional conditioning practices. *Sciences des Aliments* **13**: 89-96.

Jansen, R. C. 1993. Interval mapping of multiple quantitative trait loci. *Genetics* **135**: 205-211.

Jensen, P. and Barton-Gade, P. A. 1985. Performance and carcass characteristics of pigs with known genotypes for halothane susceptibility. In: Stress susceptibility and meat quality in pigs. *Proceedings of the Commission on Animal Management and Health and Commission of Pig Production.* EAAP publ. no. 33. p 80. Wageningen, The Netherlands.

Johnson, P. L., McEwan, J. C., Dodds, K. G., Purchas, R. W. and Blair, H. T. 2005a. A directed search in the region of GDF8 for quantitative trait loci affecting carcass traits in Texel sheep. *Journal of Animal Science* **83**:1988-2000.

Johnson, P. L., McEwan, J. C., Dodds, K. G., Purchas, R. W. and Blair, H. T. 2005b. Meat quality traits were unaffected by a quantitative trait locus affecting leg composition traits in Texel sheep. *Journal of Animal Science* **83**: 2729-2735.

Jones, H. E., Lewis, R. M., Young, M. J. and Simm, G. 2004. Genetic parameters for carcass composition and muscularity in sheep measured by X-ray computer tomography, ultrasound and dissection. *Livestock Production Science* **90**: 167-179.

Jones, H. E., Lewis, R. M., Young, M. J. and Wolf, B. T. 2002. The use of X-ray computer tomography for measuring the muscularity of live sheep. *Animal Science* **75**: 387-399.

Jones, S. D. M. 1995. Future directions in carcass assessment. pp215-228. In *Quality and grading of carcasses of meat animals.* Ed. Jones, S.D.M. CRC Press: Boca Raton.

Jones, S. D. M., Cliplef, R. L., Fortin, A. F., McKay, R. M., Murray, A. C., Pommier, S. A., Sather, A. P. and Schaefer, A. L. 1994. Production and ante-mortem factors influencing pork quality. *Pig News Info* **15** (1): 15N.

Jonsäll, A., Johansson, L. and Lundström, K. 2001. Sensory quality and cooking loss of ham muscle (M. biceps et femoris) from pigs reared indoors and outdoors. *Meat Science* **57**: 245-250.

Jopson, N. B., McEwan, J. C., Dodds, K. G. and Young, M. J. 1995. Economic benefits of including computer tomography measurements in sheep breeding programmes. *Proceedings of Australian Association for Animal Breeding and Genetics* **12**: 72-76.

Jopson, N. B., McEwan, J. C., Fennessy, P. F., Dodds, K. G., Nicoll, G. B. and Wade, C. M. 1997. Economic benefits of including computed tomography measurements in a large terminal sire breeding programme, *Proceedings of the Association for the Advancement of Animal Breeding and Genetics* **12**: 72-76.

Josell, Å., Enfält, A. C., von Seth, G., Lindahl, G., Hedebro-Velander, I., Andersson, L. and Lundström, K. 2003. The influence of RN genotype, including the new V199I allele, on the eating quality of pork loin. *Meat Science* **65**: 1341-1351.

Judd, J. T., D. J. Baer, B. A. Clevidence, P. Kris-Etherton, R. A. Muesing and M. Iwane. 2002. Dietary cis and trans monounsaturated and saturated FA and plasma lipids and lipoproteins in men. *Lipids* **37**: 123-131.

Jul, M. and Zeuthen. 1981. Quality of pig meat for fresh consumption. *Progress in Food and Nutrition Science.* 6th edition. Pergamon Press. Oxford. 350 pp.

Just, A., Pedersen, O. K., Jorgensen, H. and Kruse, V. 1986. Heritability estimates of some blood characters and intramuscular fat and their relationship with some production characters in pigs. *Pig News Info* **7**: 198.

Kanner, J. 1994. Oxidative processes in meat and meat products: Quality implications. *Meat Science* **36**: 169-189.

Keele J. W., Shackelford, S. D., Kappes, S. M., Koohmaraie, M. and Stone, R. T. 1999. A region on bovine chromosome 15 influences beef longissimus tenderness in steers. *Journal of Animal Science* **77**: 1364–1371.

Kemp, J. D., Mahyuddin, M., Ely, D. G., Fox, J. D. and Moody, W. G. 1981. Effect of feeding systems, slaughter weight and sex on organoleptic properties, and fatty acid composition of lamb. *Journal of Animal Science* **51**: 321-330.

Kempster, A. J., Cook, G. L. and Grantley-Smith, M. 1986. National estimates of the body-composition of British cattle, sheep and pigs with special reference to trends in fatness - A review. *Meat Science* **17**: 107-138.

Kempster, A. J., Cuthbertson, A. and Harrington, G. 1982. *Carcass evaluation in livestock breeding, production and marketing.* Granada: London. 306p.

Kepler, C. R., Hirons, K. P., McNeill, J. J. and Tove, S. B. 1966. Intermediates and products of the biohydrogenation of linoleic acid by butyrivibrio fibrisolvens. *Journal of Biological Chemistry* **241**: 1350-1354.

Kim, J. J., Farnir, F., Savell, J. and Taylor, J. F. 2003. Detection of quantitative trait loci for growth and beef carcass fatness traits in a cross between Bos taurus (Angus) and Bos indicus (Brahman) cattle. *Journal of Animal Science* **81**: 1933-1942.

Kirton, A. H., Woods, E. G. and Duganzich, D. M. 1983. Comparison of well and poorly muscled lamb carcasses as selected by experienced meat industry personnel. *Proceedings of the New Zealand Society of Animal Production* **43**: 111- 113.

Klont, R. E., Hulsegge, B., Hoving-Bolink, A. H., Gerritzen, M. A., Kurt, E., Winkelman-Goedhart, H. A., de Jong, I. C. and Kranen, R. W. 2001. Relationships between behavioural and meat quality characteristics of pigs raised under barren and enriched housing conditions. *Journal of Animal Science* **79**: 2835-2843.

Kmiec, M. 1999. Transferrin polymorphism versus growth rate in lambs, polish long-wool sheep: I. Frequency of genes and genotypes of transferrin in flock of polish long-wool sheep. *Archiv fur Tierzucht-Archives of Animal Breeding* **42**: 393-402.

Knott, S. A., Elsen, J. M. and Haley, C. S. 1996. Methods for multiple-marker mapping of quantitative trait loci in half-sib populations. *Theoretical and Applied Genetics* **93**: 71-80.

Knott, S. A., Marklund, L., Haley, C. S., Andersson, K., Davies, W., Ellegren, H., Fredholm, M., Hansson, I., Hoyheim, B., Lundstrom, K., Moller, M. and Andersson, L. 1998. Multiple marker mapping of quantitative trait loci in a cross between outbred wild boar and Large White pigs. *Genetics* **149**: 1069-1080.

Kok F. J. and Kromhout, D. 2004. Atherosclerosis-epidemiological studies on the health effects of a Mediterranean diet. *European Journal of Nutrition* **43**(Suppl 1): 2-5.

Kouba, M., Enser, M., Whittington, F. M., Nute, G. R. and Wood, J. D. 2003. Effect of a high-linolenic acid diet on lipogenic enzyme activities, fatty acid composition, and meat quality in the growing pig. *Journal of Animal Science* **81**: 1967-1979.

Kramer, J. K. G., Sehat, N., Dugan, M. E. R., Mossoba, M. M., Yurawecz, M. P. , Roach, J. A. G., Eulitz, K., Aalhus, J. L., Schaefer, A. L. and Ku, Y. 1998. Distributions of conjugated linoleic acid (CLA) isomers in tissue lipid classes of pigs fed a commercial CLA mixture determined by gas chromatography and silver ion high-performance liquid chromatography. *Lipids* **33:** 549-558.

Kris-Etherton P. M., Deer, J., Mitchell, D. C., Mustad, V. A., Russell, M. E., McDennell, E. T., Slabsky, D. and Pearson, T. A. 1993. The role of fatty acids saturation on plasma lipids, lipoproteins: I. Effects of whole food diets high in cocoa butter, olive oil, soybean oil, dairy butter, and milk chocolate on the plasma lipids of young men. *Metabolism* **42:** 121-129.

Kristensen, L., Therkildsen, M., Riis, B., Sorensen, M. T., Oksbjerg, N., Purslow, P. P. and Ertbjerg, P. 2002. Dietary-induced changes of muscle growth rate in pigs: Effects on *in vivo* and *post-mortem* muscle proteolysis and meat quality. *Journal of Animal Science* **80:** 2862-2871.

Kruglyak, L. and Lander, E. S. 1995. A nonparametric approach for mapping quantitative trait loci. *Genetics* **139:** 1421-1428.

Kuchtik, J., Subrt, J. and Horak, F. 1996. Quality analysis of meat of slaughter lambs. *Zivocisna Vyroba* **41:** 183-188.

Labuza, T. P. 1971. Kinetics of lipid oxidation in foods. *CRC Critical Reviews of Food Technology* **2:** 355-404.

Lai, L., Kang, J. X., Li, R., Wang, J., Witt, W., Yong, H. Y., Hao, Y., Wax, D., Murphy, C. N., Rieke, A., Samuel, M., Linville, M.L., Korte, S. W., Evans, R., Starzl, T. E., Prather, R. S. and Dai, Y. 2006. Generation of cloned transgenic pigs rich in omega-3 fatty acids. *Nature Biotechnology* **24:** 435–437.

Lambe, N. R., Young, M. J., McLean, K. A., Conington, J. and Simm, G. 2003. Prediction of total body tissue weights in Scottish Blackface ewes using computed tomography scanning. *Animal Science* **76:** 191-197.

Lander, E. S. and Botstein, D. 1989. Mapping Mendelian factors underlying quantitative traits using RFLP linkage maps. *Genetics* **121:** 185-199.

Larick, D. K. and Turner, B. E. 1990. Flavor characteristics of forage-fed and grain-fed beef as influenced by phospholipid and fatty-acid compositional differences. *Journal of Food Science* **55:** 312.

Larzul, C., Lefaucheur, L., Ecolan, P., Gogue, J., Talmant, A., Sellier, P., Le Roy, P. and Monin, G. 1997. Phenotypic and genetic parameters for longissimus muscle fiber characteristics in relation to growth, carcass, and meat quality traits in large white pigs. *Journal of Animal Science* **75**: 3126-3137.

Lauridsen, J. 1998. Stochastic simulation of alternative breeding schemes for Danish meat type sheep and ultrasonic eyemuscle and backfat measurements on lambs. *PhD Thesis.* The Royal Veterinary and Agricultural University, Copenhagen.

Lawrence, T. L. J. and V. R. Fowler. 1997. Growth of Farm Animals. CAB International, NY.

Lawrie, R. 1992. Conversion of muscle to meat. In D. E. Johnston, M. K. Knight and D. A. Ledward, *The chemistry of muscle based foods.* Royal Society of Chemistry. Cambridge. 43-61 pp.

Le Roy, P., J. Naveau, J. M. Elsen and P. Sellier. 1990. Evidence for a new major gene influencing meat quality in pigs. *Genetical Research* **55**: 33 40.

Ledward, D. A. 1992. Colour of raw and cooked meat. In J. D.E., M. K. Knight and D. A. Ledward, *The chemistry of muscle based foods.* Royal Society of Chemistry. Cambridge. 128-144 pp.

Lee, G. J., Archibald, A. L., Law, A. S., Lloyd, S., Wood, J. D. and Haley, C. S. 2004. Detection of quantitative trait loci for androstenone, skatole and boar taint in a cross between Large White and Meishan pigs. *Animal Genetics* **36:** 14-22.

Lee, K. N., Kritchevsky, D. and Pariza, M. W. 1994. Conjugated linoleic-acid and atherosclerosis in rabbits. *Atherosclerosis* **108:** 19-25.

Legrand, I., Denoyelle, C. and Quilichini, Y. 1995. Effects of breed and rationing concnentrates on lambs subcutaneous adipose tissue quality. In: *Proceedings of the 41st Annual International Congress of Meat Science and Technology* **11**: 116-117.

Lewis, R. M. and Simm, G. 2002. Small ruminant breeding programmes for meat: progress and prospects. *Proceedings of the 7th World Congress on Genetics Applied to Livestock production,* Communication no. 02-01.

Lewis, R. M., Simm, G. and Warkup, C. C. 1993. Enjoying the taste of lamb. *Meat Focus International* **2:** 393-395.

Lewis, R. M., Simm, G., Dingwall, W. S. and Murphy, S. V. 1996. Selection for lean growth in terminal sire sheep to produce leaner crossbred progeny. *Animal Science* **63**: 133-142.

Lindén, A., Andersson, K. and Oskarsson, A. 2001. Cadmium in organic and conventional pig production. *Archives of Environmental Contamination and Toxicology* **40**: 425-431.

Liu, Q., Lanari, M.C. and Schaefer, D. M. 1995. A review of dietary vitamin E supplementation for improvement of beef quality. *Journal of Animal Science* **73**: 3131-3140.

Liu, Z., Lirette, A., Fairfull, R. W. and McBride, B.W. 1994. Embryonic adenosine triphosphate: phosphate diesters ratios obtained with in vivo nuclear magnetic resonance spectroscopy (Phosphorus-31): a new technique for selecting leaner broiler chickens. *Poultry Science* **73**: 1633-1 641.

Lo, L. L. 1990. A genetic analysis of growth, real-time ultrasound, carcass and pork quality data in Duroc and Landrace swine. *MSc Thesis*. University of Illinois.

Lo, L. L., McLaren, D. G., McKeith, F. K., Fernando, R. L. and Novakofski, J. 1992. Genetic analyses of growth, real-time ultrasound, carcass, and pork quality traits in Duroc and Landrace pigs: II. Heritabilities and correlations. *Journal of Animal Science* **70**: 2387-2396.

Lundström, K., Andersson, A. and Hansson, I. 1996. Effect of the RN gene on technological and sensory meat quality in crossbred pigs with Hampshire as terminal sire. *Meat Science* **42**: 145-153.

Lundström, K., Enfält, A. C., Tornberg, E. and Agerhem, H. 1998. Sensory and technological meat quality in carriers and noncarriers of the RN allele in Hampshire crosses and in purebred Yorkshire pigs. *Meat Science* **48**:115-124.

MacDougall, D. B. 1982. Changes in the colour and opacity of meat. *Food Chemistry* **9**: 75-88.

MacDougall, D. B. and Rhodes, D. N. 1972. Characteristics of Appearance of Meat. 3. Studies on color of meat from young bulls. *Journal of the Science of Food and Agriculture* **23**: 637-647.

MacNeil, M. D. and Grosz, M. D. 2002. Genome-wide scans for QTL affecting carcass traits in Hereford x composite double backcross populations. *Journal of Animal Science* **80:** 2316-2324.

Maddox, J. F., Davies, K. P., Crawford, A. M., Hulme, D. J., Vaiman, D., Cribiu, E. P., Freking, B. A., Beh, K. J., Cockett, N. E., Kang, N., Riffkin, C. D., Drinkwater, R., Moore, S. S., Dodds, K. G., Lumsden, J. M., van Stijn, T. C., Phua, S. H., Adelson, D. L., Burkin, H. R., Broom, J. E., Buitkamp, J., Cambridge, L., Cushwa, W. T., Gerard, E., Galloway, S. M., Harrison, B., Hawken, R. J., Hiendleder, S., Henry, H. M., Medrano, J. F., Paterson, K. A., Schibler, L., Stone, R. T. and van Hest, B. 2001. An enhanced linkage map of the sheep genome comprising more than 1000 loci. *Genome Research* **11:** 1275-1289.

Mallows, C. L. 1973. Some comments on Cp. *Technometrics* **15:** 661-675.

Malmfors, B. and Nilsson, R. 1979. Meat Quality Traits in Swedish Landrace and Yorkshire Pigs with Special Emphasis on Genetics. *Acta Agriculturae Scandinavica* 81-90.

Mariani, P., Lundström, K., Gustafsson, U., Enfält, A. C., Juneja, R. K. and Andersson, L. 1996. A major locus (RN) affecting muscle glycogen content is located on pig chromosome 15. *Mammalian Genome* **7:** 52-54.

Mason, P. 2004. Fatty acids: which ones do we need? *The Pharmaceutical Journal* **273:** 750-752.

Mattson, F. H. and Grundy, S. M. 1985. Comparison of effects of dietary saturated, monounsaturated, and polyunsaturated fatty acids on plasma lipids and lipoproteins in man. *Journal of Lipid Research* **26:** 194-202.

McEwan, J. C., Fennessy, P. F., Greer, G. J., Bruce, G. D. and Bain, W. E. Selection for carcass composition: effects on the development of subcutaneous and internal fat depots. *Proceedings of the Nutrition Society of New Zealand* **14:** 163-164.

McGeehin, B., Sheridan, J. J. and Butler, F. 2001. Factors affecting the pH decline in lamb after slaughter. *Meat Science* **58:** 79-84.

McRae, A. F., Bishop, S. C., Walling, G. A., Wilson, A. D. and Visscher, P. M. 2005. Mapping of multiple quantitative trait loci for growth and carcass traits in a complex commercial sheep pedigree. *Animal Science* **80:** 135-141.

Meat and Livestock Commission 1998. *Sheep yearbook.* Meat and Livestock Commission, Milton Keynes, UK.

Mehrabian, M., Allayee, H., Stockton, J., Lum, P. Y., Drake, T. A., Castellani, L. W., Suh, M., Armour, C., Edwards, S., Lamb, J., Lusis, A. J. and Schadt, E. E. 2005. Integrating genotypic and expression data in a segregating mouse population to identify 5-lipoxygenase as a susceptibility gene for obesity and bone traits. *Nature Genetics* **37**: 1224-1233.

Mendenhall, V. T. and S. K. Ercanbrack. 1979. Effect of carcass weight, sex, and breed on consumer acceptance of lamb. *Journal of Food Science* **44**: 1063–1066.

Meuwissen, T. H. E. and Goddard, M. E. 1996. The use of marker haplotypes in animal breeding schemes. *Genetics Selection Evolution* **28**: 161-176.

Milan, D., Le Roy, P., Woloszyn, N., Caritez, J. C., Elsen, J. M., Sellier, P. and Gellin, J. 1995. The RN locus for meat quality maps to pig chromosome 15. *Genetic Selection Evolution* **27**:195-199.

Miller, D. C. 1996. Accuracy and application of real-time ultrasound for evaluation of carcasses' merit in live animals. *Animal Husbandry Newsletter, May 1996.*

Mitchell, A. D. Wang, P. C. Rosebrough, R. W. Elsasser, T. H. and Schmidt, W. F., 1991. Assessment of body composition of poultry by nuclear magnetic resonance imaging and spectroscopy. *Poultry Science,* **70**: 2494-2500.

Moloney, A. P., Keane, M. G., Dunne, P. G., Mooney, D. T. and Trot, D. J. 2001. Delayed concentrate feeding in a grass silage/concentrate beef finishing system: effects on fat colour and meat quality. *Proceedings of the 47th International Conference of Meat Science and Technology*, Krakow, Poland.

Monin, G., Brard, C., Vernin, P. and Naveau, J. 1992. Effects of the RN   gene on some traits of muscle and liver in pigs. In: *Proceedings of the 38th International Congress on Meat Science Technology* **3**: 391-394.

Moody, D. E., Pomp, D., Newman, S. and MacNeil, M. D. 1996. Characterization of DNA polymorphisms in three populations of Hereford cattle and their associations with growth and maternal EPD in line 1 Herefords. *Journal of Animal Science* **74**: 1784-1793.

Moser, D. W., Bertrand, J. K., Misztal, I., Kriese, L. A. and Benyshek, L. L. 1998. Genetic parameter estimates for carcass and yearling ultrasound measurements in Brangus cattle. *Journal of Animal Science* **76**: 2542-2548.

Mottram, D. S. 1998. Flavour formation in meat and meat products: a review. *Food Chemistry* **62**: 415-424.

Mrode, R. A., Smith, C. and Thompson, R. 1990. Selection for rate and efficiency of lean gain in Hereford cattle. 1. Selection pressure applied and direct responses. *Animal Production* **51**: 23-34.

Mulvihill, B. 2001. Ruminant meat as a source of conjugated linoleic acid (CLA). *Nutrition Bulletin* **26**: 295-299.

National Cholesterol Education Program. 2001. Executive Summary of the Third Report of the National Cholesterol Education Program (NCEP) Expert Panel on Detection, Evaluation, and Treatment of High Blood Cholesterol in Adults (Adult Treatment Panel III). *Circulation* **285**: 2486–97.

Naveau, J. 1986. Contribution á l'étude du de´terminisme génétique de la qualité de la viande porciné. Héritabilité du rendement technologique Napole. *Journal Recherche Porcine France* **18**: 265-276.

Naveau, J., Pommeret, P. and Leschaux, P. 1985. Proposition d'une méthode de mesure du rendement technologique: la "methode Napole". *Techni-Porcine* **8**:713.

Nephawe, K. A., Cundiff, L. V., Dikeman, M. E., Crouse, J. D., and Van Vleck, L. D. 2004. Genetic relationships between sex-specific traits in beef cattle: Mature weight, weight adjusted for body condition score, height and body condition score of cows, and carcass traits of their steer relatives. *Journal of Animal Science* **82**: 647-653.

Nicoll, G. B., Burkin, H. R., Broad, T. E., Jopson, N. B., Greer, G. J., Bain, W. E., Wright, C. S., Dodds, K. G., Fennessy, P. F. and McEwan, J. C. 1998. Genetic linkage of microsatellite markers to the Carwell locus for rib-eye muscling in sheep. *Proceedings of the 6th World Congress on Genetics Applied to Livestock Production, Armidale, Australia* **26**: 529-532.

Nicoll, G. B., McEwan, J. C., Dodds, K. G. and Jopson, N. B. 1997. Genetic improvement in Landcorp Lamb Supreme terminal sire flocks. *Association for the Advancement of Animal Breeding and Genetics* **12**: 68-71.

Nicolosi, R. J., Rogers, E. J., Kritchevsky, D., Scimeca, J. A. and Huth, P. J. 1997. Dietary conjugated linoleic acid reduces plasma lipoproteins and early aortic atherosclerosis in hypercholesterolemic hamsters. *Artery* **22**:266-277.

Nürnberg, K., Dannenberger, D., Nuernberg, G., Ender, K., Voigt, J., Scollan, N. D., Wood, J. D., Nute, G. R. and Richardson, R. I. 2005. Effect of a grass-based and a concentrate feeding system on meat quality characteristics and fatty acid composition of longissimus muscle in different cattle breeds. *Livestock Production Science* **94**: 137-147.

Offer, G. and Knight, M. K. 1988. The structural basis of water-holding in meat. Part 2: Drip losses. In R. Lawrie, *Developments in meat science 4*. Elsevier Applied Science Publication. London. 172-234 pp.

Offer, G., Knight, P., Jeacocke, R., Almond, R., Cousins, T., Elsey, J., Parsons, N., Sharp, A., Starr, R. and Purslow, P. P. 1989. The structural basis of the water-holding, appearance and toughness of meat and meat products. *Food Microstructure* **8**: 151-170.

O'Halloran, G. R., Troy, D. J., Buckley, D. J. and Reville, W. J. 1997. The role of endogenous proteases in the tenderisation of fast glycolysing muscle. *Meat Science* **47**: 187-210.

Okuyama H., Kobayashi T. and Watanabe S. 1997. Dietary fatty acids-the n-6/n-3 balance and chronic elderly diseases. Excess linoleic acid and relative n-3 deficiency syndrome seen in Japan. *Prog. Lipid. Res.* **35**: 409-457

Olson, L. W., Dickerson, G. E. and Glimp, H. A. 1976. Selection criteria for intensive market lamb production: growth traits. *Journal of Animal Science* **43**: 78 89.

Pariacote, F., Van Vleck, L. D. and Hunsley, R. E. 1998. Genetic and phenotypic parameters for carcass traits of American Shorthorn beef cattle. *Journal of Animal Science* **76**: 2584-2588.

Perez-Enciso, M., Clop, A., Noguera, J. L., Ovilo, C., Coll, A., Folch, J. M., Babot, D., Estany, J., Oliver, M. A., Diaz, I. and Sanchez, A. 2000. A QTL on pig chromosome 4 affects fatty acid metabolism: Evidence from an Iberian by Landrace intercross. *Journal of Animal Science* **78**: 2525-2531.

Pommier, S. A., Houde, A., Rousseau, F. and Savoie, Y. 1992. The effect of the malignant hyperthermia genotype as determined by a restriction endonuclease assay on carcass characteristics of commercial crossbred pigs. *Canadian Journal of Animal Science* **72**: 973.

Pré, J. 1991. Lipid peroxidation. *Pathology Biology* **39**: 716–736.

Price, M. A. Development of carcass grading and classification systems. 1995. Future directions in carcass assessment. pp173-199. In *Quality and grading of carcasses of meat animals*. Ed. Jones, S.D.M. CRC Press: Boca Raton.

Puntila, M. L., Mäki, K. and Rintala, O. 2002. Assessment of carcass composition based on ultrasonic measurements and EUROP conformation class of live lambs. *Journal of Animal Breeding and Genetics* **119:** 367-378.

Quali, A. 1992. Proteolytic and physicocohemical mechanisms involved in meat texture development. *Biochimie* **74:** 251-265.

Quintanilla, R., Demeure, O., Bidanel, J. P., Milan, D., Iannuccelli, N., Amigues, Y., Gruand, J., Renard, C., Chevalet, C. and Bonneau, M. 2003. Detection of quantitative trait loci for fat androstenone levels in pigs. *Journal of Animal Science* **81:** 385-394.

Renand, G., Jurie, C., Robelin, J., Picard, B., Geay, Y. and Ménissier, F. 1995. Genetic variability of muscle biological characteristics of young Limousin bulls. *Genetic Selection Evolution* **27:** 287.

Rhee, K. S. 1992. Fatty acids in meats and meat products. Pages 65–93 in *Fatty Acids in Foods and Their Health Implications.* C. K. Chow, ed. Marcel Dekker, New York.

Richardson, R. I., Nute, G. R., Wood, J. D., Scollan. N. D. and Warren, H. E. 2004. Effects of breed, diet and age on shelf life, muscle vitamin E and eating quality of beef. In: *Proceedings of the British Society of Animal Science,* York, p.84.

Risvik, E. 1994. Sensory properties and preferences. *Meat Science* **36:** 67-77.

Rivellese, A. A., Maffettone, A., Vessby, B., Uusitupe, M., Hermansen, K., Berglund, L., Loureranta, A., Meyer, B. J. and Riccardi, G. 2003. Effects of dietary saturated, monounsaturated and n-3 fatty acids on fasting lipoproteins, LDL size and post-prandial lipid metabolism in healthy subjects. *Atherosclerosis* **167:** 149-58.

Roden, J. A., Merrell, B. G., Murray, A. W. and Haresign, W. 2003. Genetic analysis of live weight and ultrasonic fat and muscle traits in a hill sheep flock undergoing breed improvement utilizing an embryo transfer programme. *Animal Science* **76:** 367-373.

Romans, J. R., Costello, W. J., Carlson, C. W., Greaser, M. L. and Jones, K. W. 1994. The meat we eat. Interstate Publishers, Danville, Illinois.

Rosenvold, K. and Andersen, H. J. 2003. Factors of significance, for pork quality - a review. *Meat Science* **64:** 219-237.

Rubino, R., Morand-Fehr, P., Renieri, C., Peraza, C. and Sarti, F. M. 1999. Typical products of the small ruminant sector and the factors affecting their quality. *Small Ruminant Research* **34:** 289-302.

Saatci, M., Ap Dewi, I. and Ulutas, Z. 1998. Correlations between weaning weight, carcass traits and ultrasonic body measurements in Welsh Mountain sheep. *Proceedings of the British Society of Animal Science*, p.84.

Sañudo, C. and I. Sierra. 1982. A study on the carcass and meat quality in Romanov x Aragon crossbreds. I. A description and comparison of two types of lambs. *Annals Faculty Veterinary University Zaragoza* **16/17**: 285.

Sañudo, C., Enser, M. E., Campo, M. M., Nute, G. R., Maria, G., Sierra, I. and Wood, J. D. 2000. Fatty acid composition and sensory characteristics of lamb carcasses from Britain and Spain. *Meat Science* **54**: 339–346.

Sañudo, C., Nute, G. R., Campo, M. M., Maria, G., Baker, A., Sierra, I., Enser, M. E. and Wood, J. D. 1998. Assessment of commercial lamb meat quality by British and Spanish taste panels. *Meat Science* **48**: 91-100.

SAS 2003. *SAS/STAT user's guide. Version 9.1.3* SAS Institute Inc., Cary, NC.

Sather, A. P., Jones, S. D. M. and Tong, A. K. W. 1991a. Halothane genotype by weight interactions on lean yield from pork carcasses. *Canadian Journal of Animal Science* **71**: 633.

Sather, A. P., Jones, S. D. M., Tong, A. K. W. and Murray, A. C. 1991b. Halothane genotype by weight interactions on pig meat quality. *Canadian Journal of Animal Science* **71**: 645.

Sather, A. P., Murray, A. C., Zawadski, S. M. and Johnson, P. 1991c. The effect of the halothane gene on pork production and meat quality of pigs reared under commercial conditions. *Canadian Journal of Animal Science.* **71**: 959.

Schäfer, A., Rosenvold, K., Purslow, P. P., Andersen, H. J. and Henckel, P. 2002. Physiological and structural events postmortem of importance for drip loss in pork. *Meat Science* **61**: 355-366.

Scheper, J. 1979. Influence of Environmental and Genetic-Factors on Meat Quality. *Acta Agriculturae Scandinavica* 20-31.

Scholz, A., Baulain, U. and Kallweit, E. 1993. Quantitative analyse von bildern aus der Magnet-Resonanz-Tomographie. *Zuchtungskunde* **65**: 206-215.

Schönfeld, H. 2001. Consumer trends in South Africa with reference to meat and pork. Lecture presented at the Mini Symposium. *The role of pork in the New Millenium.* ARC-Irene.

Schrooten, C. and Visscher, A. H. 1987. Genetic parameters for growth and slaughter quality. IVO-DLO Report B-283. Instituut voor Veeteeltkundig Onderzoek Schoonoord Zeist, The Netherlands.

Schworer, D., Morel, P., Prabucki, A. and A. Rebsamen. 1987. Selection for intramuscular fat in pigs. Results of investigations in Switzerland. *Die Tierzuechter* **39:** 392.

Scislowski, V., Bauchart, D., Gruffat, D., Laplaud, P. M. and Durand, D. 2005. Effects of dietary n-6 or n-3 polyunsaturated fatty acids protected or not against ruminal hydrogenation on plasma lipids and their susceptibility to peroxidation in fattening steers. *Journal of Animal Science* **83:** 2162-2174.

Seaton, G., Haley, C. S., Knott, S. A., Kearsey, M. and Visscher, P. M. 2002. QTL Express: mapping quantitative trait loci in simple and complex pedigrees. *Bioinformatics* **18:** 339-340.

Sehested, E. 1984. Computerised tomography in sheep. In *In vivo measurement of body composition in meat animals* (ed. D Lister), pp. 67-74. Elsevier, London.

Sellier, P. 1988. Genetics of meat and carcass traits. In: *The genetics of the pig* (ed. MF Rothschild and A Ruvinsky), pp. 463-510. CAB International, Wallingford.

Sellier, P. and Monin, G. 1994. Genetics of pig meat quality: A review. *Journal of Muscle Foods* **5**.

Simm, G. 1994. Developments in improvement of meat sheep. *Proceedings of the 5th World Congress on Genetics Applied to Livestock Production*, vol. **18**, pp. 3–10.

Simm, G. and Dingwall, W. S. 1989. Selection indices for lean meat production in sheep. *Livestock Production Science* **21:** 223-233.

Simm, G., Lewis, R. M., Grundy, B. and Dingwall, W. S. 2002. Responses to selection for lean growth in sheep. *Animal Science* **74:** 39-50.

Simm, G., Young, M. J. and Beatson, P. R. 1987. An economical selection index for lean meat production in New Zealand sheep. *Animal Production* **45:** 465-475.

Simopoulos, A. P. 2001. n-3 fatty acids and human health: Defining strategies for public policy. *Lipids* **36** (Suppl.)**:**83-89.

Simpson, S. P. and A. J. Webb. 1989. Growth and carcass performance of British Landrace pigs heterozygous at the halothane locus. *Animal Production* **49:** 503.

Skjervold, H., Gronseth, K., Vangen, O. and Evensen, A. 1981. In vivo estimation of body composition by computerized tomography. *Journal of Animal Breeding and Genetics* **98**: 77-79.

Slater, T. F. 1984. Free-radical mechanisms in tissue injury. *Biochemistry Journal* **222**: 1– 15.

Spanier, A. M., Flores, M., McMillan, K. W. and Bidner, T.D. 1997. The effect of post-mortem aging on meat flavour quality in Brangus beef. Correlation of treatments, sensory, instrumental and chemical descriptors. *Food Chemistry* **59**: 531-538.

Spelman, R. J., Coppieters, W., Karim, L., van-Arendonk, J. A. M. and Bovenhuis, H. 1996. Quantitative trait loci analysis for five milk production traits on chromosome six in the Dutch Holstein-Friesian population. *Genetics* **144**: 1799-1808.

Splan, R. K., Cundiff, L. V. and Van Vleck, L. D. 1998. Genetic parameters for sex-specific traits in beef cattle. *Journal of Animal Science* **76**: 2272-2278.

Stanford, K., Jones, S. D. M., and Price, M. A. 1998. Methods of predicting lamb carcass composition: A review. *Small Ruminant Research* **29**: 241-254.

Stone, R. T., Keele, J. W., Shackelford, S. D., Kappes, S. M. and Koohmaraie, M. 1999. A primary screen of the bovine genome for quantitative trait loci affecting carcass and growth traits. *Journal of Animal Science* **77**: 1379-1384.

Streitz, E., Baulain, U. and Kallweit, E., 1995. Investigation of body composition of growing lambs by means of magnetic resonance imaging. Zuchtungskunde, **67**: 392-403.

Swatland, H. J. 1995. On-Line Evaluation of Meat, *Technomic*, Lancaster, PA. pp. 347.

Szabo, C., Babinszky, L., Verstegen, M. W. A., Vangen, O., Jasman, A. J. M. and Kanis, E. 1999. The application of digital imaging techniques in the in vivo estimation of the body composition of pigs: a review. *Livestock Production Science* **60**: 1-11.

Taylor, St. C. S. 1985. Use of genetic size-scaling in evaluation of animal growth. *Journal of Animal Science* **61** (Suppl. 2): 118-143.

Temple, R. S., Stonaker, H. H., Hovry, D., Posakony, G. and Hazaleus, M. H. 1956. Ultrasonic and conductivity methods for estimating fat thickness in live cattle. *Proceedings of the Western Section, American Society of Animal Science* **7**: 477-481.

Thorsteinsson, S. and Eythorsdottir, E. Genetic parameters of ultrasonic and carcass cross-sectional measurements and muscle and fat weight of Icelandic lambs. *Proceedings of the 6th World Congress on Genetics Applied to Livestock Production* **24**: 149-152.

Touraille, C. 1992. Reducing fat in meat animals. *Meat Science* **31:** 243-244.

Van Laack, R., Stevens, S. G. and Stalder, K. J. 2001. The influence of ultimate pH and intramuscular fat content on pork tenderness and tenderization. *Journal of Animal Science* **79:** 392-397.

Van Oeckel, M. J., Warnants, N. and Boucque, C. V. 1999. Pork tenderness estimation by taste panel, Warner-Bratzler shear force and on-line methods. *Meat Science* **53:** 259-267.

Van Trijp, H., Steenkamp, J.-B. and Candel, M. 1997. Quality labelling as instrument to create product equity: The case of IKB in the Netherlands. In: Wierenga, B., van Tilburg, A., Grunert, K., Steenkamp, J.-B. and Wedel, M. (eds), *Agricultural Marketing and Consumer Behaviour in a Changing World* (pp. 201-216). Dordrecht: Kluwer Academic Publishers.

Van Vleck, L. D., Hakim, A. F., Cundiff, L. V., Koch, R. M., Crouse, J. D. and Boldman, K. G. 1992. Estimated breeding values for meat characteristics of crossbred cattle with an animal model. *Journal of Animal Science* **70:** 363-371.

Vance, D. E. and Vance, J. 1996. Biochemistry of Lipids, Lipoproteins and Membranes. Elsevier, The Hague.

Vangen, O. 1988. In *World Conference on Animal Production*, 12p.

Vangen, O. and Jopson, N.B. 1996. Research application of non-invasive techniques for body composition. *Proceedings of the 47th annual meeting of the European Association for Animal Production, Lillehammer,* Norway, pp. 25–29.

Vangen, O. and Skjervold, H. 1981. Estimating body composition in pigs by computerized tomography. *Pig news and information* **2:**153-154.

Vangen, O., Kvame, T., Haugen, S., Avdem, F. and Eikje, S. 2003. Use of a meat sheep sire line to improve product quality in a national sheep breeding system. *Proceedings of the 54th Annual Meeting of the European Association for Animal Production,* Rome, Italy.

Vilkki, H. J., De Koning, D. J., Elo, K., Velmala, R. and Maki-Tanila, A. 1997. Multiple marker mapping of quantitative trait loci of Finnish dairy cattle by regression. *Journal of Dairy Science* **80:** 198-204.

Visscher, P. M., Thompson, R. and Haley, C. S. 1996. Confidence intervals in QTL mapping by bootstrapping. *Genetics* **143:** 1013-1020.

Wachira, A. M., Sinclair, L. A., Wilkinson, R. G., Enser, M., Wood, J. D. and Fisher, A. V. 2002. Effects of dietary fat source and breed on the carcass composition, n-3 polyunsaturated fatty acid and conjugated linoleic acid content of sheep meat and adipose tissue. *British Journal of Nutrition* **88**: 697-709.

Walling, G. A., Archibald, A. L., Cattermole, J. A., Downing, A. C., Finlayson, H. A., Nicholson, D., Visscher, P. M., Walker, C. A. and C. S. Haley. 1998a. Mapping of quantitative trait loci on porcine chromosome 4. *Animal Genetics* **29**: 415-424.

Walling, G. A., Archibald, A. L., Visscher, P. M. and Haley, C. S. 1998b. Mapping genes for growth rate and fatness in a Large White x Meishan F_2 pig population. *Proceedings of the British Society of Animal Science,* 7.

Walling, G. A., Haley, C. S., Perez-Enciso, M., Thompson, R. and Visscher, P. M. 2002. On the mapping of QTLs at marker and non-marker locations. *Genetical Research* **79**: 97-106.

Walling, G. A., Visscher, P. M. and Haley, C. S. 1998b. A comparison of bootstrap methods to construct confidence intervals in QTL mapping. *Genetical Research* **71**: 171-180.

Walling, G. A., Visscher, P. M., Simm, G. and Bishop, S. C. 2001. Confirmed linkage for QTLs affecting muscling in Texel sheep on chromosome 2 and 18. *52nd Annual Meeting of the European Association for Animal Production, Budapest* 59.

Walling, G. A., Visscher, P. M., Wilson, A. D., McTeir, B. L., Simm, G. and Bishop, S. C. 2004. Mapping of quantitative trait loci for growth and carcass traits in commercial sheep populations. *Journal of Animal Science* **82**: 2234-2245.

Ward, C. E., Trent, A. and Hildebrand, J. L. 1995. Consumer perception of lamb compared with other meats. *Journal of Sheep and Goat Research* **11**: 64-70.

Warriss, P. D., Bevis, E. A. and Ekins, P. J. 1989. The relationships between glycogen stores and muscle ultimate pH in commercially slaughtered pigs. *British Veterinary Journal* **145**: 378-383.

Watanabe, A., Daly, C.C. and Devine, C. E. 1996. The effects of the ultimate pH of meat on tenderness changes during ageing. *Meat Science,* **42**: 67-78.

Webb, A. J., Carden, A. E., Smith, C. and Imlah, P. 1982. Porcine stress syndrome in pig breeding. *Proceedings of 2nd World Congress on Genetics Applied to Livestock Production* **5**: 588. Madrid, Spain.

Webb, E. C. and N. H. Casey. 1995. Genetic differences in fatty acid composition of subcutaneous adipose tissue in Dorper and SA Mutton Merino wethers at different live weights. *Small Ruminant Research* **18:** 81–88.

West, D. B., Delany, J. P., Camet, P. M., Blohm, F., Truett, A. A. and Scimeca, J. 1998. Effects of conjugated linoleic acid on body fat and energy metabolism in the mouse. *American Journal of Physiology (Regul. Integr.Comp. Phys.)* **275:** R667-R672.

Whittington, F. M., Prescott, N. J., Wood, J. D. and Enser, M. 1986. The effect of dietary linoleic-acid on the firmness of backfat in pigs of 85 kg live weight. *Journal of Science Food and Agriculture* **37:** 753-761.

Williams, C. M. 2000. Dietary fatty acids and human health. *Annales de Zootechnie* **49:** 165-180.

Wilson, L. L., Mccurley, J. R., Ziegler, J. H. and Watkins, J. L. 1976. Genetic parameters of live and carcass characters from progeny of polled Hereford sires and Angus-Holstein cows. *Journal of Animal Science* **43:** 569-576.

with eating quality. *Meat Science* **29:** 295-307.

Wood, J. D. 1990. Consequences for meat quality of reducing carcass fatness. In *Reducing fat in meat animals* (ed. J. D. Wood and A. V. Fisher), pp. 344-397. Elsevier Applied Science, London.

Wood, J. D. and Enser, M. 1997. Factors influencing fatty acids in meat and the role of antioxidants in improving meat quality. *British Journal of Nutrition* **78:** S49-S60.

Wood, J. D. and Fisher, A.V. 1990. Improving the quality of lamb meat-taste, fatness and consumer appeal. New developments in sheep production, *British Society of Animal Production.* Occasional Publication No.**14:** 99-107.

Wood, J. D. and N. D. Cameron. 1994. Genetics of meat quality in pigs. *Proceedings of the 5th World Congress on Genetics Applied to Livestock Production* **19:** 458-464.

Wood, J. D., Brown, S. N., Nute, G. R., Whittington, F. M., Perry, A. M., Johnson, S. P. and Enser, M. 1996. Effects of breed, feed level and conditioning time on the tenderness of pork. *Meat Science* **44:** 105-112.

Wood, J. D., Enser, M., Fisher, A. V., Nute, G. R., Richardson, R. I. and Sheard, P. R. 1999. Manipulating meat quality and composition. *Proceedings of the Nutrition Society* **58:** 363-370.

Wood, J. D., Enser, M., Fisher, A. V., Nute, G. R., Richardson, R. I. and Sheard, P. R. 1998. Meat quality: an integrated approach for the future. Pages 103-113 in *Proceedings of the 15th Congress of the International Pig Veterinary Society*, Nottingham.

Wood, J. D., Enser, M., Fisher, A. V., Nute, G. R., Richardson, R. I. and Sheard, P. R. 1999. Manipulating meat quality and composition. *Proceedings of Nutrition Society* **58**: 363-370.

Wood, J. D., Nute, G. R., Richardson, R. I., Whittington, F. M., Southwood, O., Plastow, G., Mansbridge, R., da Costa, N. and Chang, K. C. 2004a. Effects of breed, diet and muscle on fat deposition and eating quality in pigs. *Meat Science* **67**: 651-667.

Wood, J. D., Richardson, R. I., Nute, G. R., Fisher, A. V., Campo, M. M., Kasapidou, E., Sheard, P. R. and Enser, M. 2004. Effects of fatty acids on meat quality: a review. *Meat Science* **66**: 21-32.

Wood, J. D., Warriss, P. D. and Enser, M. B. 1992. Effects of production factors on meat quality in pigs. In D. E. Johnston and M. K. Knight, *The chemistry of muscle-based foods*. Royal Society of Chemistry. London. 3-14 pp.

Wood, J. D., Wiseman, J. and D. J. A. Cole. 1994. *Control and manipulation of meat quality*. Pages 433-456 in Principles of Pig Science. D. J. A. Cole, J. Wiseman, and M. A. Varley, Nottingham University Press, Nottingham.

Yokoyama, M. T., Carlson, J. R. and Dickinson, E. O. 1975. Ruminal and plasma concentrations of 3-methylindole associated with tryptophan-induced pulmonary edema and emphysema in cattle. *American Journal of Veterinary Research* **36**: 1349-1352.

Young, M. J. 1989. The influence of changes in tissue shape on muscle: bone ratio in growing sheep. *Proceedings of the British Society of Animal Production*. In. *Animal production* **48**: 635.

Young, M. J. and Sykes, A. R. 1987. Bone growth and muscularity. *Proceedings of the New Zealand Society of Animal Production* **47**: 73-75.

Young, M. J., Garden, K. L. and Knopp, T.C. 1987. Computer aided tomography-comprehensive body compositional data from live animals. *Proceedings of the New Zealand Society of Animal Production* **47**: 69-71.

Young, M. J., Lewis, R. M., McLean, K. A., Robson, N. A. A., Fraser, J., Fitzsimons, J., Donbavand, J. and Simm, G. 1999. Prediction of carcass composition in meat breeds of sheep using computer tomography. *Proceedings of the British Society of Animal Science* **43**.

Young, M. J., Logan, C. M., Beatson, P. R. and Nsoso, S. J. 1996. Prediction of carcass tissue weight *in vivo* using live weight, ultrasound or X-ray CT measurements,*Proceedings of the New Zealand Society of Animal Production* **56**: 205-211.

Young, M.J., Simm, G. and Glasbey, C.A. 2001. Computerised tomography for carcass analysis. *Proceedings of the British Society of Animal Science*, 250–254.

Young, O. A., Berdague, J.-L., Viallon, C., Rousset-Akrim, S. and Theriez, M. 1997. Fat-borne volatiles and sheepmeat odour. *Meat Science* **45:** 183-200.

Young, O. A., Braggins, T. J., West, J. and Lane, G. A. 1999. Animal production origins of some meat color and flavor attributes. In *Quality attributes of muscle foods* (ed. Y. L. Xiong, C. T. Ho, and F. Shahidi), pp. 11-30, Kluver Academic/Plenum Publishers, New York.

Young, O. A., Lane, G. A., Priolo, A. and Fraser, K. 2003. Pastoral and species flavor in lambs raised on pasture, lucerne or maize. *Journal of the Science of Food and Agriculture* **83:** 93-104.

Young, O. A., Reid, D. H. and Scales, G. H. 1993. Effects of breed and ultimate pH on the odour and flavour of sheep meat. *New Zealand Journal of Agricultural Research* **36**: 363-370.

Zeng, Z. B. 1993. Theoretical basis for separation of multiple linked gene effects in mapping quantitative trait loci. *Proceedings of the National Academy of Sciences of the United States of America* **90:** 10972-10976.

Zock, P. L. and Katan, M. B. 1992. Hydrogenation alternatives - Effects of trans-fatty-acids and stearic-acid versus linoleic-acid on serum-lipids and lipoproteins in humans. *Journal of Lipid Research* **33:** 399-410.

VDM publishing house ltd.

Scientific Publishing House

offers

free of charge publication

of current academic research papers, Bachelor´s Theses, Master's Theses, Dissertations or Scientific Monographs

If you have written a thesis which satisfies high content as well as formal demands, and you are interested in a remunerated publication of your work, please send an e-mail with some initial information about yourself and your work to *info@vdm-publishing-house.com*.

Our editorial office will get in touch with you shortly.

VDM Publishing House Ltd.
Meldrum Court 17.
Beau Bassin
Mauritius
www.vdm-publishing-house.com

Made in the USA
Monee, IL
14 February 2023

27657570R00108